KU-761-106

NAME THAT PLANT

AN ILLUSTRATED GUIDE TO PLANT AND BOTANICAL LATIN NAMES

FOREWORD BY
MARTIN PAGE

WORTH PRESS

First published in 2001 by

Worth Press Limited
Cambridge, England
www.worthpress.co.uk

This edition published 2006

ISBN-10: 1 903025 26 5
ISBN-13: 978-1-903025-26-0

© Worth Press Ltd, 2001, 2006

CIP catalogue records for this book are available from
the British Library and the Library of Congress.

All rights reserved. No part of this publication may be
reproduced or used in any form or by any means –
photographic, electronic or mechanical, including
photocopying, recording, taping or information
storage retrieval systems – without the written
permission of the publisher.

Designed and produced by
David Porteous Editions
for Worth Press Ltd

Typeset in Times

Printed in Finland by WS Bookwell

COUNTY STORE

SEP 15
1 7 AUG 2017

HAMPSHIRE COUNTY LIBRARY

OUT OF PRINT

WITHDRAWN

HELD AT

LAST COPY STORE

HAMPSHIRE COUNTY LIBRARY

WITHDRAWN.

581.014

NAME THAT...

Get **more** out of libraries

Please return or renew this item by the last date shown.
You can renew online at www.hants.gov.uk/library
Or by phoning 0845 603 5631

Hampshire
County Council

C013953541

CONTENTS

HAMPSHIRE COUNTY LIBRARY	
C013953541	
HJ	11/09/2006
R581.014	£6.99
1903025265	

FOREWORD

I have always thought it slightly ironic that the greatest legacy of the Roman Empire has been its art and language. By contrast the military conquests of Rome have long ago passed into history.

It could be argued that Latin was the first international language. As the Roman Empire spread the colonised peoples adopted Latin as their formal language. Latin survived the fall of Rome because it became the official language of the Roman Catholic Church and during the Middle Ages Latin was the preferred language of educated people throughout Western Europe.

The scientific study of plants dates back to the sixteenth-century, and early descriptions were, not surprisingly, made in Latin. In *Paradisi in Sole Paradisus Terrestris* by John Parkinson (1567-1650) described the female peony as *Paeonia femina vulgaris flore simplici*, which means the ordinary single female peony. While this system did work it was incredibly cumbersome.

The person who changed everything was the Swedish botanist Carl Linnaeus (1707-1778). Until Linnaeus the method of naming plants varied and it was difficult for botanists to determine whether they were referring to the same plant. Linnaeus's system used two Latin words to describe each species. The first name referred to the genus, while the second indicated the species. This combination of names is referred to as the binomial system. The principal of scientific priority starts with Linnaeus and the oldest valid scientific names are those listed in his *Species Plantarum*, which was published in 1753.

For most gardeners Latin names appear very confusing but once you understand the basic rules they are actually rather easy to use. It is quite common for people to refer to plants by either their common or vernacular names. However, the problem is that some genera have hundreds of species, and some common species may have dozens of vernacular names. A good example of this is *Caltha palustris*, which is called 'Bachelor's Buttons' in Somerset, 'Horse-Blob' in Surrey and 'Water Gowlan' in Cumbria. These names will be quite meaningless to someone from Germany or France, but they will probably know what *Caltha palustris* is.

Some names are quite common. If you see a plant with the specific name *palustris* (growing in marshes) you can be pretty certain that it will do well in wet soil, while if the name is *sylvatica* (of, or relating to, woods) it will like shady conditions. Latin names should be seen as a form of shorthand – don't be frightened of them because they can be very useful.

Martin Page
May 2001

Martin Page studied botany at the University of Wales and received his PhD from the University of Exeter. His work on the plant ecology of hay meadows was incorporated into the National Vegetation Classification.

More recently he has studied the taxonomy of the genus *Paeonia* and his first book, *The Gardener's Guide to Growing Peonies*, was published in1997. He currently works in book publishing as a Horticultural and Photographic Editor. He is now writing his second book on peonies.

PRONUNCIATION

The aim of this book is not to encourage readers to try to emulate what may be termed a classical pronunciation of Latin but simply to help them understand and pronounce botanical Latin names with a degree of ease and fluency. The overall objective is quite simply to be understood. With this in mind, phonetic spelling has been adopted so that each syllable can be spoken as it is spelt. However, this system is by no means infallible because the pronunciation may well be influenced by either common usage or overtones of regional accents, both of which may alter the vowel sounds and the emphasis.

The phonetic spelling used is based on the widespread convention of pronouncing botanical names as if they were English, but it is important to stress the right syllable and vowel sounds.

The syllable to be stressed is always shown in italic. With two-syllable words the emphasis is on the first syllable but with longer Latin words the emphasis is normally on the penultimate syllable.

Each vowel is pronounced separately, so some syllables may contain a single vowel. The dipthongs (ae and oe) are regarded as single syllables. The stressed (italic) vowel is normally long when on its own or when following a consonant (*a* as in 'father', *e* as in 'feet', *o* as in 'note'). It is short when between two consonants or when preceding a consonant (*a* as in 'pat', *e* as in 'pet', *o* as in 'pot'). The u is generally an ew sound, as in 'cue'.

The consonants c and g may be soft (as the *c* in 'ace' or the *g* in 'gel') or hard (as the *c* in 'cat' or the *g* in 'got').

GLOSSARY

acuminate
Tapered to a long, narrow point.

anther
The part of the stamen bearing the pollen, forming the male part of the flower.

areole
A cluster of spines (in cacti).

auricle
A lobe, shaped like an ear.

awn
A bristle growing from the tip of flowering parts of certain grasses and cereals.

bi-pinnate
Doubly pinnate, i.e. leaflets, which grow opposite each other in a feather shape, have leaflets also arranged in this way.

bract
A leaf-like structure or scale beneath a flower or group of flowers. This can be brightly-coloured, and make more visual impact than the flower (eg Poinsettia, Bougainvillea).

carpel
Part of the female portion of a flower. This consists of an ovary (or part of one), style and stigma.

column
A long structure in an orchid, consisting of the combined stamens and style.

compound
Made up of several parts.

corolla
Floral envelope of a flower, made up of the petals.

corymb
A flat-topped or convex flower cluster forming an inflorescence where the outer flowers open first.

dioecious
With male and female reproductive organs borne in separate flowers on separate plants.

disc (or disk)
A flattened, round part of a plant, eg a composite flower receptacle or the middle part of an orchid lip.

entire
With a smooth margin: not lobed or toothed.

epiphytic
Growing on another plant but not parasitically.

filament
The stalk of the stamen, bearing the anther at its tip.

glabrous
Smooth, without hairs.

glaucous
With a whitish, blue-green, or grey powdery bloom.

graft-hybrid
A plant formed when one plant is grafted onto a different one. A different plant is created, though the genes are not mixed.

indusium
The membrane covering the sorus (spore case) in ferns.

inflorescence
The part of the plant consisting of flower-bearing stalks.

involucre
A ring of bracts at the base of a flower or inflorescence.

linear
Narrow and elongated, with parallel sides.

lip
A modified petal (eg in orchids or the Mint family).

monocarpic
Dying after flowering and fruiting only once.

mucronate
Ending suddenly, with a short, abrupt point.

node
A point on a plant stem from which leaves or lateral branches grow.

obovate
Shaped like the cross-section of an egg with the broader end above.

ovate
Shaped like the cross-section of an egg with the narrower end above.

panicle
A raceme with the branches branched (eg the oat).

pappus
A tuft of fine, feathery hairs or bristles on the fruit.

pedicel
The stalk of a single flower in an inflorescence.

peltate
A leaf with the stalk attached in the centre of the lower surface and not on the margin.

perianth
Outer part of a flower comprising the petals and sepals, often used when they cannot easily be differentiated.

petiole
The stalk of the leaf.

phyllode
A flattened leaf-stalk which resembles and functions as a leaf.

pinna
The leaflet of a pinnate compound leaf.

pinnule
The secondary division of a bi-pinnate leaf.

pollinium
A mass of pollen grains transported as a whole during pollination.

proliferous
reproducing vegetatively.

pseudobulb
The thickened stem of an orchid.

raceme
An inflorescence bearing single, stalked flowers from a central axis, the youngest at the apex.

receptacle
The apex of a flower stem on which the floral parts are borne.

scape
A leafless flower stalk

sepal
The parts of a perianth outside of the petals.

sessile
Unstalked.

simple

BIBLIOGRAPHY

For readers who wish to delve further into this fascinating subject the publications listed below are highly recommended for further reading.

The Internet, with its ever-increasing range of web-sites relating to all aspects of plant life, has also provided a rich source of information in the preparation of this book and will reward those who wish to 'surf'.

Botanical Latin
William T. Stearn
David & Charles

Plant Names Simplified
A.T. Johnson and H.A. Smith
Landsmans Bookshop Ltd

A Gardener's Handbook of Plant Names
A.W. Smith
Dover Publications

Dictionary of British and Irish Botanists and Horticulturists
R.Desmond
Taylor and Francis

Dictionary of English Plant Names
G. Grigson
Allen Lane

LATIN-ENGLISH PLANT NAMES

A

Abelia, a-*bel*-ee-a. *Caprifoliaceae.*
After Dr Clarke Abel. Deciduous,
semi-evergreen and evergreen shrubs.
  chinensis, chin-*en*-sis. Of China.
  floribunda, flo-ri-*bun*-da. Profusely
  flowering.
  grandiflora, gran-di-*flo*-ra. Large-
  flowered.
  schumannii, shoo-*man*-ee-ee. After
  K. M. Schumann.
  triflora, tri-*flo*-ra. Three-flowered.

Abeliophyllum, a-bel-ee-o-*fil*-lum.
Oleaceae. From *Abelia* and Gk. *phyllon* (leaf). Deciduous shrub.
  distichum, dis-tik-um. In two ranks
  (leaves). White Forsythia.

Abies, *a*-bee-eez. *Pinaceae.* Classical
L. name. Evergreen conifers. Fir.
  alba, al-ba. White (bark). Silver Fir.
  amabilis, a-*ma*-bi-lis. Beautiful. Red
  Silver Fir.
  balsamea, bal-*sam*-ee-a. Balsam-
  producing (bark). Balsam Fir.
  cephalonica, sef-a-*lon*-i-ka. Of
  Cephalonia. Greek Fir.
  concolor, kon-ko-lor. Of the same
  colour. White Fir.
  delavayi, del-a-*vay*-ee. After Abbé
  Jean Marie Delavay.
  grandis, gran-dis. Large. Giant Fir.
  homolepis, ho-mo-*lep*-is. With similar scales. Nikko Fir.
  koreana, Ko-ree-*a*-na. Of Korea.
  procera, pro-*see*-ra. Tall. Noble Fir,
  Christmas Tree.
  veitchii, veech-ee-ee. After the
  Veitch nursery.

Abronia, a-*bro*-nee-a. *Nyctaginaceae.*
From Gk. *abros* (delicate), the bracts.

Annual and perennial herbs.
  latifolia, la-ti-*fo*-lee-a. Broad-leaved.
  Yellow Sand Verbena.
  umbellata, um-bel-*la*-ta. With
  umbels. Pink Sand Verbena.

Abutilon, a-*bew*-ti-lon. *Malvaceae.*
From the Arabic for a mallow-like
plant. Tender, semi-evergreen and
deciduous shrubs. Flowering Maple.
  globosum, glo-*bo*-sum. Spherical
  x *hybridum, hib*-rid-um. Hybrid.
  Chinese Lantern.
  megapotamicum, meg-a-pot-*am*-ik-
  um. Of the big river (Rio Grande,
  Brazil).
  pictum, pik-tum. Painted (flowers).
  striatum, stri-*a*-tum. Striped.
  x *suntense,* sun-*ten*-see. From Sunte
  House, Sussex.
  vitifolium, vee-ti-*fo*-lee-um. *Vitis*-
  leaved.

Acacia, a-*kay*-she-a. *Leguminosae.*
Gk. name from *akis* (sharp point), after
the thorns. Tender or semi-hardy,
deciduous and evergreen trees and
shrubs. Mimosa, Wattle.
  baileyana, bay-lee-*a*-na. After F. M.
  Bailey. Golden Mimosa.
  cultriformis, kul-tree-*form*-is. Knife-
  shaped. Knife-leaf Wattle.
  dealbata, dee-al-*ba*-ta. Whitened.
  Mimosa.
  longifolia, long-i-*fo*-lee-a. With long
  leaves. Sallow Wattle.
  podalyriifolia, pod-a-li-ree-i-*fo*-lee-a.
  With *Podalyria*-like leaves.
  pravissima, pra-*vis*-im-a. Very
  crooked. Oven's Wattle.
  verticillata, ver-ti-sil-*a*-ta. Whorled.
  Prickly Moses.

13

Acaena, a-*see*-na. *Rosaceae*. From Gk. *akaina* (thorn). Perennial herbs and sub-shrubs. New Zealand Burr.
  buchananii, bew-kan-*an*-ee-ee. After John Buchanan.
  microphylla, mie-kro-*fil*-a. Small-leaved.

Acalypha, a-ka-*lee*-fa, *Euphorbiaceae*. From Gk. *akelpe* (nettle). Tender, ever-green, perennial herbs and shrubs.
  hispida, his-pid-a. Bristly. Red Hot Cat's Tail.
  wilkesiana, wilks-ee-*a*-na. After Admiral Charles Wilkes. Jacob's Coat.

Acantholimon, a-kanth-o-*lee*-mon. *Plumbaginaceae*. From Gk. *akanthos* (thorn) and *limonium* (sea lavender). Evergreen perennial herbs.
  glumaceum, gloo-ma-*see*-um. Having glumes (bracts enclosing the flowers).
  venustum, ven-*us*-tum. Charming.

Acanthus, a-*kanth*-us. *Acanthaceae*. From Gk. *akanthos* (thorn). Perennial herbs. Bear's Breeches.
  hungaricus, hun-*ga*-ri-kus. Of Hungary.
  longifolius, long-i-*fo*-lee-us. Long-leaved.
  mollis, mol-lis. Soft.
  spinosus, spi-*no*-sus. Spiny (leaves).

Acer, *ay*-ser, *Aceraceae*. L. *name*. Deciduous or evergreen trees and shrubs. Maple.
  cappadocicum, kap-a-*do*-kik-um. Of Cappadocia. Caucasian Maple.
  carpinifolium, kar-pie-ni-*fo*-lee-um. *Carpinus*-leaved. Hornbeam Maple.
  circinatum, ser-sin-*a*-tum. Rounded (leaves). Vine Maple.
  crataegifolium, kra-tee-gi-*fo*-lee-um.

Crataegus-leaved. Hawthorn Maple.
  davidii, da-*vid*-ee-ee. After Abbé Armand David.
  ginnala, jin-*na*-la. Native name.
  griseum, gris-ee-um. Grey. Paper Bark Maple.
  grosseri, gro-se-ree. After Grosser.
  japonicum, ja-*pon*-i-kum. Of Japan. Japanese Maple.
  lobelii, lo-*bel*-ee-ee. After Mathias de l'Obel.
  macrophyllum, mak-ro-*fil*-um. Large-leaved. Oregon Maple.
  monspessulanum, mon-spes-ew-*la*-num. Of Montpelier. Montpelier Maple.
  negundo, ne-*gun*-do. Native name. Ash-leaved Maple. Box Elder.
  nikoense, nik-o-*en*-see. Of Nikko, Japan.
  palmatum, pal-*ma*-tum. Hand-like (leaves).
  pensylvanicum, pen-sil-*van*-i-kum. Of Pennsylvania. Striped Maple.
  platanoides, pla-ta-*noi*-deez. Like *Platanus*. Norway Maple.
  pseudoplatanus, sood-o-*pla*-ta-nus.

Acer pseudoplatanus

False *Platanus*. Sycamore, Great
Maple.
rubrum, rub-rum. Red. Red Maple.
rufinerve, roof-i-*ner*-vee. Red-
veined.
saccharinum, sak-a-*ree*-num. Sugary
(sap). Silver Maple.
saccharum, sa-*ka*-rum. Sugar cane
(Gk. *sacchoron),* produces maple
syrup. Sugar Maple.

Achillea, a-*kil*-lee-a. *Compositae.*
After Achilles. Semi-evergreen perennial herbs.
clavennae, kla-*ven*-ee. After Niccola
Chiavera.
clypeolata, kli-pee-o-*la*-ta. Shield-
shaped (flower head).
depressa, dee-*pres*-sa. Flattened.
filipendulina, fi-li-pen-dew-*lee*-na.
Filipendula-like.
millefolium, mil-lee-*fo*-lee-um.
Thousand-leaved. Yarrow.
ptarmica, tar-mi-ka. Sneezing plant
(Gk. *ptarmos*). Sneezewort.
taygetea, tay-*gee*-tee-a. From the
Taygetos mountains, Greece.

Achillea ptarmica

Achimenes, a-kim-*ee*-neez.
Gesneriaceae. From Gk. *cheimino*
(suffer from cold). Tender, trailing
perennials. Hot-water Plant.
erecta, e-*rek*-ta. Erect.
grandiflora, gran-di-*flo*-ra. Large-
flowered.

Acidanthera, a-sid-an-*the*-ra.
Iridaceae. From Gk. *akis* (point) and
anthera (anther). Tender or semi-
hardy, cormous perennial herbs.
bicolor, bi-kol-or. Two-coloured
(flowers).
murielae, mew-ree-*el*-ee. After
Muriel.

Acokanthera, a-ko-kan-*the*-ra.
Apocynaceae. From Gk. *akoke* (point)
and *anthera* (anther). Tender, evergreen trees and shrubs. Poison Tree.
oblongifolia, ob-long-i-*fo*-lee-a.
With oblong leaves.
spectabilis, spek-*ta*-bi-lis.
Spectacular.

Aconitum, a-kon-*ie*-tum.
Ranunculaceae. L. name. Poisonous
perennial herbs. Monkshood.
x *bicolor, bi*-kol-or. Two-coloured
(flowers).
carmichaelii, kar-mie-*keel*-ee-ee.
After J. R. Carmichael.
napellus, na-*pel*-lus. A small turnip.
Garden Monkshood.
volubile, vol-*ew*-bi-lee. Twining.
vulparia, vul-*pa*-ree-a. Of Wolves.
Wolf's Bane.

Acorus, a-*ko*-rus. *Araceae.* L. name.
Semi-evergreen and perennial aquatic
herbs.
calamus, ka-la-mus. Reed-like.
Myrtle Flag.
gramineus, gra-*min*-ee-us. Grass-
like.

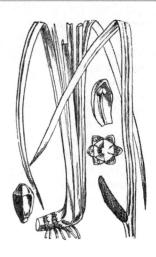

Acorus calamus

Acradenia, ak-ra-*deen*-ee-a. *Rutaceae.*
From Gk. *akros* (at the tip) and *adenia*
(gland). Semi-hardy, evergreen shrubs.
   frankliniae, frank-*lin*-ee-ee. After
   Lady Franklin. Whitey Wood.

Actaea, ak-*tee*-a. *Ranunculaceae.*
From Gk. *aktea* (elder). Perennial
herbs with poisonous fruits. Baneberry.
   alba, al-ba. White. White Baneberry.
   rubra, rub-ra. Red. Red Baneberry.

Actinidia, ak-tin-*id*-ee-a.
Actinidiaceae. From Gk. *aktinos* (ray).
Deciduous climbers.
   chinensis, chin-*en*-sis. Of China.
   Chinese Gooseberry, Kiwi Fruit.
   kolomikta, ko-lo-*mik*-ta. Native name.
   polygama, po-*lig*-a-ma. Polygamous
   (mixed gender flowers). Silver Vine.

Adenophora, a-den-*o*-fo-ra.
Campanulaceae. From Gk. *aden*
(gland) and *phorea* (bear). Perennial
herbs.
   lilifolia, lil-ee-i-*fo*-lee-a. *Lilium*-
   leaved

Adiantum, a-dee-*an*-tum.
Adiantaceae. From Gk. *adiantos* (dry).
Tender deciduous and semi-evergreen
ferns. Maidenhair Fern.
   capillus-veneris, ka-*pil*-lus-*ven*-e-ris.
   Venus's hair, the delicate foliage.
   Common Maidenhair.
   caudatum, kaw-*da*-tum. With a tail.
   Trailing Maidenhair.
   pedatum, ped-*a*-tum. Like a bird's
   foot (fronds). Five-fingered
   Maidenhair Fern.
   raddianum, ra-dee-*a*-num. After
   Giuseppe Raddi. Delta Maidenhair
   Fern.
   venustum, ven-*us*-tum. Handsome.
   Evergreen Maidenhair.

Adiantum capillus-veneris

Adonis, a-*do*-nis. *Ranunculaceae.*
After Adonis, the Gk. god. Perennial
herbs.
   amurensis, am-oor-*en*-sis. Of the
   Amur river region.
   vernalis, ver-*na*-lis. Of spring.

Adromischus, a-dro-*mis*-kus.
Crassulaceae. From Gk. *hadros* (stout)

mischos (stalk). Tender, perennial succulents and evergreen sub-shrubs.
  maculatus, mak-ew-*la*-tus. Spotted
  (leaves). Calico Hearts.

Aechmea, eek-*mee*-a. *Bromeliaceae.*
From Gk. *aichme* (point). Tender,
evergreen perennial herbs.
  fasciata, fa-see-*a*-ta. Banded
  (leaves). Urn Plant.
  fulgens, ful-jenz. Shining (bracts).
  recurvata, re-*kur*-va-ta. Curved
  downwards

Aeonium, ee-*o*-nee-um. *Crassulaceae.*
The L. name of one species. Tender or
semi-hardy succulents.
  arboreum, ar-*bor*-ee-um. Tree-like.
  canariense, ka-nar-ee-*en*-see. Of the
  Canary Islands. Velvet Rose.

Aerides, air-ee-deez. *Orchidaceae.*
From Gk. *aer* (air). Tender, epiphytic
orchids. Fox-tail Orchid.
  crispa, kris-pa. Finely waved.
  falcata, fa-*ka*-ta. Sickle-shaped
  (leaves).
  odorata, o-do-*ra*-ta. Scented.

Aeschynanthus, ee-skee-*nan*-thus.
Gesneriaceae. From Gk. *aischune*
(shame) and *anthos* (flower). Tender
evergreen climbers or creeping perennial herbs.
  pulcher, pul-ker. Pretty. Lipstick
  Plant.
  speciosus, spes-ee-*o*-sus. Showy.

Aesculus, *es*-ku-lus.
Hippocastanaceae. L. for an oak with
edible acorns. Deciduous trees and
shrubs. Horse Chestnut.
  californica, kal-i-*forn*-i-ka. Of
  California. California Buckeye.
  x *carnea, kar*-nee-a. Flesh-coloured
  (flowers). Red Horse-chestnut.

flava, fla-va. Yellow (flowers).
Yellow Buckeye.
  hippocastanum, hip-o-*ka*-sta-num. L.
  name. Horse Chestnut.
  indica, in-di-ka. Of India. Indian
  Horse Chestnut.
  parviflora, par-vi-*flo*-ra. Small-flowered.

Aethionema, eeth-ee-o-*nee*-ma.
Cruciferae. From Gk. *aitho* (scorch)
and *nema* (thread), colour of the stamens. Perennials and sub-shrubs.
Stone Cress.
  armenum, ar-*meen*-um. Of Armenia.
  grandiflorum, gran-di-*flo*-rum.
  Large-flowered.
  iberideum, i-be-*ri*-dee-um. *Iberis*-
  like.
  pulchellum, pul-*kel*-lum. Pretty.

Agapanthus, ag-a-*panth*-us. *Liliaceae.*
From Gk. *agape* (love) and *anthos*
(flower). Hardy and semi-hardy perennial herbs.
  africanus, af-ri-*ka*-nus. African.
  African Lily.
  campanulatus, kam-pan-ew-*la*-tus.
  Bell-shaped (flowers).
  inapertus, in-a-*per*-tus. Closed.
  orientalis, o-ree-en-*ta*-lis. From the
  east.
  praecox, pree-kox. Early (flowering).

Agapetes, ag-a-*peet*-eez. *Ericaceae.*
From Gk. *agapetos* (desirable).
Evergreen, or deciduous semi-hardy
shrubs and climbers.
  affinis, a-*fee*-nis. Related to.
  macrantha, ma-*kranth*-a. Large-
  flowered.
  rugosa, roo-*go*-sa. Wrinkled
  serpens, ser-penz. Creeping.

Agastache, a-*ga*-sta-kee. *Labiatae.*
From Gk. *agan* (very much) and

stachys (spike). Perennial herbs.
  mexicana, mex-i-*ka*-na. Of Mexico.

Agave, a-*ga*-vee. *Agavaceae.* From
Gk. *agave* (noble). Tender and semi-
hardy perennial succulents. Century
Plant.
  americana, a-me-ri-*ka*-na. Of
  America.
  attenuata, a-ten-ew-*a*-ta. Drawn out.
  filifera, fil-i-fe-ra. Bearing threads
  (leaf margins). Thread Agave.
  parviflora, par-vi-*flo*-ra. Small-flowered.
  victoriae-reginae, vik-*tor*-ree-ie-ree-
  jeen-ee. After Queen Victoria.

Ageratum, aj-er-*a*-tum. *Compositae.*
From Gk. *a* (not) and *geras* (age).
Annual and biennial herbs. Floss
Flower.
  houstonianum, hew-ston-ee-*a*-num.
  After William Houston.
  mexicanum, mex-i-*ka*-num.
  Mexican.

Aglaonema, a-gla-o-*nee*-ma. *Araceae.*
From Gk. *aglaos* (bright) and *nema*
(thread). Tender, evergreen perennial
herbs.
  commutatum, kom-ew-*ta*-tum.
  Changeable.
  costatum, kos-*ta*-tum. Ribbed.
  Spotted Evergreen.
  crispum, kris-pum. Finely wavy.
  pictum, pik-tum. Painted (variegated
  leaves).

Agrostemma, ag-ro-*stem*-a.
Caryophyllaceae. From Gk. *agros*
(field) and *stemma* (garland). Annual
herbs.
  coeli-rosea, see-lee-*ro*-see-a. Rose of
  Heaven.
  githago, gi-*tha*-go. L. name. Corn
  Cockle.

Agrostemma githago

Aichryson, ie-*kris*-on. *Crassulaceae.*
Gk. name. Tender annual and perennial succulents.
  x *domesticum,* dom-*es*-ti-kum.
  Cultivated.

Ailanthus, ie-*lan*-thus.
Simaroubaceae. From *ailanthos* (tree
of heaven), the Moluccan name.
Deciduous tree.
  altissima, al-*tis*-si-ma. Tallest. Tree
  of Heaven.

Ajuga, a-*joo*-ga. *Labiatae.* From Gk. *a*
(not) and *zeugon* (yoke). Annual and
perennial herbs.
  alpina, al-*pie*-na. Alpine.
  australis, aw-*stra*-lis. Southern.
  genevensis, gen-e-*ven*-sis. Of
  Geneva. Blue Bugle.
  pyramidalis, pi-ra-mid-*a*-lis.
  Pyramidal.
  reptans, rep-tanz. Creeping. Bugle.

Akebia, a-*kee*-bee-a. *Lardizabalaceae.*
From the Japanese name. Deciduous

or semi-evergreen climbers.
quinata, kwi-*na*-ta. In fives
(leaflets).
trifoliata, tri-fo-lee-*a*-ta. With three
leaves.

Alchemilla, al-ke-*mil*-la. *Rosaceae.*
From Arabic *alkemelych* (alchemy).
Perennial herbs. Lady's Mantle.
alpina, al-*pie*-na. Alpine. Alpine
Lady's Mantle.
conjuncta, kon-*junk*-ta. Joined.
mollis, mol-lis. Softly hairy (leaves).
Lady's Mantle

Alisma, a-*lis*-ma. *Alismataceae.* Gk.
name. Deciduous, perennial aquatic
herbs.
natans, na-tanz. Floating.
plantago-aquatica, plan-*ta*-go-a-
kwa-ti-ka. Water plantain.

Allamanda, al-a-*man*-da,
Apocynaceae. After Dr. Frederick
Allamand. Tender, evergreen climbers.
cathartica, ka-*thar*-ti-ka. Purging.
Golden Trumpet.

Ajuga genevensis

Alchemilla alpina

Allium, *a*-lee-um. *Liliaceae.* L. name
for garlic. Perennial herbs.
aflatunense, a-fla-tun-*en*-see. Of
Aflatun.
beesianum, beez-ee-*a*-num. After
Bees Nursery.
cepa, ce-pa. L. name. Onion,
Shallot.
cyaneum, sie-*a*-nee-um. Blue (flowers).
cyathophorum, sie-a-tho-*fo*-rum.
Cup-bearing.
farreri, fa-ra-ree. After Reginald
Farrer.
fistulosum, fis-tew-*lo*-sum. Hollow-
stemmed. Welsh Onion.
flavum, fla-vum. Yellow (flowers).
Small Yellow Onion.
giganteum, ji-*gan*-tee-um. Very
large.
karataviense, ka-ra-ta-vee-*en*-see. Of
the Kara Tau.
moly, mo-lee. Gk. name.
macranthum, ma-*kranth*-um. Large-
flowered.
narcissiflorum, nar-sis-i-*flo*-rum.
Narcissus-flowered.
neapolitanum, nee-a-pol-i-*ta*-num.
Of Naples. Daffodil Garlic.

oreophilum, o-ree-*o*-fi-lum.
Mountain-loving.
porrum, po-rum. L. name. Leek.
pulchellum, pul-*kel*-lum. Pretty.
schoenoprasum, skeen-o-*pra*-sum.
Rush-like leaves. Chives.

Alnus, *al*-nus. *Betulaceae.* L. name.
Deciduous trees and shrubs. Alder.
cordata, kor-*da*-ta. Heart-shaped
(leaves). Italian Alder.
fruticosa, froo-ti-*ko*-sa. Shrubby.
glutinosa, gloo-ti-*no*-sa. Sticky
(shoots and leaves). Common Alder.

Alnus glutinosa

incana, in-*ka*-na. White (under the
leaves). Grey Alder.
japonica, ja-*pon*-i-ka. Of Japan.

Alocasia, alo-*ka*-see-a. *Araceae.* From
Gk. *Calocasia.* Tender, evergreen,
perennial herbs. Elephant's Ear Plant.
cuprea, kew-pree-a. Coppery.
macrorrhiza, mak-ro-*ree*-za. With a
large root. Giant Taro.
odora, o-*do*-ra. Scented.
picta, pik-ta. Painted.
sanderiana, san-da-ree-*a*-na. After
the Sander nursery. Kris Plant.
veitchii, veech-ee-ee. After the
Veitch nursery.

Aloe, *a*-lo-ee. *Liliaceae.* From the
Arabic name. Tender, succulent, evergreen shrubs, perennials and climbers.
arborescens, ar-bo-*res*-enz. Tree-like.
Torch Plant.
aristata, a-ris-*ta*-ta. With a long,
bristle-like tip (leaves). Lace Aloe.
brevifolia, brev-i-*fo*-lee-a. With short
leaves.
ciliaris, si-lee-*a*-ris. Fringed with
hairs. Climbing Aloe.
ferox, fe-rox. Spiny. Cape Aloe.
humilis, hum-i-lis. Low-growing.
Spider Aloe
striata, stri-*a*-ta. Striped. Coral Aloe.
variegata, va-ree-a-*ga*-ta.
Variegated. Tiger Aloe.
vera, ve-ra. True.

Alonsoa, a-*lon*-zo-a.
Scrophulariaceae. After Alonzo
Zanoni. Semi-hardy annuals. Mask
Flower.
warscewiczii, var-sha-*vich*-ee-ee.
After Joseph Warscewicz.

Aloysia, a-lo-*is*-ee-a. *Verbenaceae.*
After Maria Louisa, Queen of Spain.
Semi-hardy, deciduous or evergreen
shrubs.
triphylla, tri-*fil*-a. With three leaves.
Lemon Verbena.

Alstroemeria, al-strurm-*e*-ree-a.
Alstroemeriaceae. After Baron Claus
Alstroemer. Semi-hardy perennial
herbs. Peruvian Lily.
aurantiaca, aw-ran-tee-*a*-ka.
Orange.
hookeri, huk-a-ree. After W. J.
Hooker.
ligtu, lig-too. Chilean name. St.
Martin's Flower.

Alternanthera, al-ter-nan-*the*-ra.
Amaranthaceae. From L. *alternans*

(alternating) and *anthera* (anther).
Tender perennial herbs.

amoena, a-*mee*-na. Pleasant.
ficoidea, fee-*koi*-dee-a. *Ficus*-like.
versicolor, ver-*si*-kol-or. Variously
coloured (leaves).

Althaea, al-*thee*-a. *Malvaceae*. From
Gk. *althaine* (heal). Annual or perennial herbs.

ficifolia, fi-ki-*fo*-lee-a. *Ficus*-leaved
officinalis, o-fis-i-*na*-lis. Sold in
shops. Marsh Mallow.
rosea, ro-see-a. Rose-coloured

Alyssum maritimum

Amaranthus, am-a-*ran*-thus.
Amaranthaceae. From Gk. *amarantos*
(unfading). Annual herbs.

caudatus, kaw-*da*-tus. With a tail.
hybridus, hib-ri-dus. Hybrid.
tricolor, tri-kol-or. Three-coloured
(leaves). Chinese Spinach.

Amaryllis, am-a-*ril*-lis.
Amaryllidaceae. After a mythological
Gk. shepherdess. Bulbous herb.

belladonna, bel-a-*don*-a. Beautiful
lady. Jersey Lily.

Althaea officinalis

Alyssum, a-*lis*-sum. *Cruciferae*. From
Gk. *a* (not) and *lyssa* (madness),
alleged to cure rabies. Annual and
perennial herbs. Madwort.

argenteum, ar-*jen*-tee-um. Silvery.
idaeum, ie-*de*-um. Of Mt. Ida.
maritimum, ma-*ri*-ti-mum. Growing
near the sea.
montanum, mon-*ta*-mum. Of mountains.
repens, ree-penz. Creeping.
serpyllifolium, ser-pil-li-*fo*-lee-um.
Thyme-leaved.

Amelanchier, a-me-*lan*-kee-er.
Rosaceae. From the French name for
A. ovalis. Deciduous trees and shrubs.
Juneberry.

alnifolia, al-ni-*fo*-lee-a. Alnus-
leaved.
arborea, ar-*bo*-ree-a. Tree-like.
canadensis, kan-a-*den*-sis. Of
Canada.
laevis, lee-vis. Smooth (leaves).
lamarckii, la-*mar*-kee-ee.
After Lamarck.

Amorpha, a-*mor*-fa. *Leguminosae.*
From Gk. *amorphos* (deformed).
Deciduous shrubs and sub-shrubs.
  canescens, ka-*nes*-enz. Greyish-
  white hairs. Lead Plant.
  fruticosa, froo-ti-*ko*-sa. Shrubby.
  False Indigo.

Amsonia, *am*-son-ee-a. *Apocynaceae.*
After Dr Charles Amson. Perennial
herbs. Blue Star.
  salicifolia, sa-li-si-*fo*-lee-a. Salix-
  leaved.
  tabernaemontani, ta-ber-nie-mon-*ta*-
  nee. After Jakob Theodor von
  Bergzabern.

Anacyclus, an-a-*sik*-lus. *Compositae.*
From Gk. *an* (without) *anthos* (a
flower) and *kuklos* (a ring). Perennial
herbs.
  depressus, dee-*pres*-sus. Flattened.

Anagallis, an-a-*ga*-lis. *Primulaceae.*
From Gk. *anagelao* (delight). Annual
and creeping perennial herbs.
Pimpernel.
  arvensis, ar-*ven*-sis. Of cultivated
  fields. Scarlet Pimpernel.
  tenella, ten-*el*-la. Dainty. Bog
  Pimpernel.

Anagallis arvensis

Ananas, *a*-na-nas. *Bromeliaceae.*
South American name. Tender, evergreen, perennial herbs. Pineapple.
  bracteatus, brak-tee-*a*-tus. With
  bracts. Wild Pineapple.
  comosus, kom-*o*-sus. With a tuft of
  leafy bracts. Pineapple.

Anaphalis, a-*na*-fa-lis. *Compositae.*
From the Gk. name for an everlasting
plant. Perennial herbs.
  margaritacea, mar-ga-ri-*ta*-see-a.
  Pearl-like (flower head). Pearly
  Everlasting.
  triplinervis, tri-plee-*ner*-vis. Three-
  veined.
  yedoensis, yed-o-*en*-sis. Of Yeddo
  (Tokyo).

Anchusa, an-*kew*-sa. *Boraginaceae.*
From Gk. *ankousa* (cosmetic paint).
Semi-hardy, annual, biennial or perennial herbs.
  azurea, a-*zew*-ree-a. Sky-blue.
  caespitosa, see-spi-*to*-sa. Tufted.
  capensis, ka-*pen*-sis. Of the Cape of
  Good Hope.

Andromeda, an-*drom*-e-da.
Ericaceae. After the mythological Gk.
princess, Andromeda. Evergreen
shrubs.
  polifolia, pol-i-*fo*-lee-a. White-
  leaved. Marsh Andromeda.

Androsace, an-*dros*-a-see.
Primulaceae. From Gk. *aner* (man)
and *sakos* (shield). Annual and evergreen perennial herbs. Rock Jasmine.
  carnea, *kar*-nee-a. Flesh-coloured.
  chamaejasme, kam-ee-*jas*-mee.
  Dwarf jasmine.
  lanuginosa, la-noo-gi-*no*-sa. Woolly.
  pyrenaica, pi-ren-*ee*-i-ka. Of the
  Pyrenees.
  sarmentosa, sar-men-*to*-sa.

Producing runners.
sempervivoides, sem-per-vee-*voi*-deez. *Sempervivum*-like.
villosa, vil-*lo*-sa. Softly hairy.

Anemone nemorosa

Andromeda polifolia

Anemone, a-*nem*-o-nee.
Ranunculaceae. From Gk. *anemos*
(wind). Perennial herbs. Windflower.
   alpina, al-*pie*-na. Alpine.
   apennina, a-pen-*nee*-na. Of the
   Apennines.
   biflora, bi-*flo*-ra. Two-flowered.
   blanda, blan-da. Pleasant.
   coronaria, ko-ro-*na*-ree-a. Used in
   garlands.
   elongata, e-long-*ga*-ta. Elongated.
   fulgens, ful-jenz. Shining.
   hupehensis, hew-pee-*hen*-sis. Of
   Hupeh, China.
   hybrida, hib-ri-da. Hybrid.
   japonica, ja-*pon*-i-ka. Of Japan.
   narcissiflora, nar-sis-i-*flo*-ra.
   Narcissus-flowered.
   nemorosa, nem-o-*ro*-sa. Of woods.
   Wood Anemone.
   rivularis, reev-ew-*la*-ris. Growing
   by streams.
   sylvestris, sil-*ves*-tris. Of woods.

Anemonopsis, a-nem-o-*nop*-sis.
Ranunculaceae. From *Anemone* and
Gk. *-opsis* (resemblance). Perennial
herb.
   macrophylla, mak-ro-*fil*-a. Large-
   leaved.

Angelica, an-*jel*-i-ka. *Umbelliferae.*
From its alleged angelic healing properties. Perennial herb.
   archangelica, ark-an-*jel*-i-ka. After
   the Archangel Raphael.

Angraecum, an-*gree*-kum.
Orchidaceae. From the Malayan
angurek (air plants). Greenhouse
orchids.
   eburneum, eb-*ur*-ne-um. Ivory-like.
   sesquipedale, ses-kwee-ped-*a*-lee.
   1½ feet long (spur). Comet Orchid.

Anguloa, an-gew-*lo*-a. *Orchidaceae.*
After Don Francisco de Angulo.
Greenhouse orchids. Cradle Orchid.
   clowesii, klowz-ee-ee. After the Rev.
   John Clowes.

Anigozanthus, a-nee-go-*zan*-thus.
Haemadoreaceae. From Gk. *anoigo*
(open) and *anthos* (flower). Kangaroo
Paw.
   flavidus, fla-vi-dus. Yellow.
   Evergreen Kangaroo Paw.

manglesii, mang-*galz*-ee-ee. After Robert Mangles.

Antennaria, an-ten-*a*-ree-a. *Compositae.* From L. antenna (ship's sail yard). Evergreen or semi-evergreen perennial herbs.
   dioica, dee-o-*ee*-ka. Dioecious. Cat's Foot.

Antennaria diocia

Anthemis, *an*-them-is. *Compositae.* The Gk. name for *Camomile.* Perennial herbs. Camomile, Dog Fennel.
   nobile, no-bi-lee. Notable.
   punctata, punk-*ta*-ta. Spotted.
   tinctoria, tink-*to*-ree-a. Used in dyeing. Dyer's Camomile, Yellow Camomile.

Anthericum, an-*the*-ri-kum. *Liliaceae.* From Gk. *antherikon* (asphodel). Perennial herbs. Spider Plant.
   liliago, lil-ee-*a*go. *Lilium*-like. St Bernard Lily.
   ramosum, ra-*mo*-sum. Branched.

Anthurium, an-*thew*-ree-um. *Araceae.* From Gk. *anthos* (flower) and *oura* (tail). Tender, evergreen perennial herbs.
   andreanum, an-dree-*a*-num. After

Edouard Francis André. Flamingo Flower.
   crystallinum, kris-tal-*lee*-num. Crystalline.
   scherzerianum, skairts-a-ree-*a*-num. After Herr Scherzer.
   veitchii, veech-ee-ee. After the Veitch nursery. King Anthurium.

Anthyllis, an-*thil*-lis. *Leguminosae.* Gk. name. Perennial herbs and shrubs.
   barba-jovis, bar-ba-*jo*-vis. Jupiter's beard.
   hermanniae, her-*ma*-nee-ee. *Hermannia*-like.
   montana, mon-*ta*-na. Of mountains.

Antigonon, an-*ti*-go-non. *Polygonaceae.* From Gk. *anti* (like) and *polygonon* (knotweed). Tender evergreen climber.
   leptopus, lep-to-pus. Slender-stalked. Coral Vine, Chain of Love.

Antirrhinum, an-tee-*rie*-num. *Scrophulariaceae.* From Gk. *anti* (like) and *rhis* (snout). Perennial herbs and

Anthemis tinctoria

Antirrhinum majus

semi-evergreen sub-shrubs.
Snapdragon.
   majus, ma-jus. Larger. Snapdragon.

Aphelandra, af-el-*an*-dra.
Acanthaceae. From Gk. *apheles* (simple) and *aner* (male). Tender evergreen shrubs and perennials..
   squarrosa, skwa-*ro*-sa. With parts
   spreading. Zebra Plant.
   louisae, loo-*eez*-ee. After Queen
   Louise of Belgium.

Apium, *a*-pee-um. *Umbelliferae.* The
L. name for celery and parsnip.
Biennial herb.
   graveolens, gra-*vee*-o-lenz. Strong
   smelling. Wild Celery.
   dulce, dul-see. Sweet. Celery.
   rapaceum, ra-*pa*-see-um. Like a
   turnip. Celeriac.

Aponogeton, a-pon-o-*jee*-ton.
Aponogetonaceae. Origin unknown.
Deciduous, perennial aquatic herbs.
   distachyos, di-*sta*-kee-os. With two
   spikes. Water Hawthorn.

Aporocactus, a-po-ro-*kak*-tus.
Cactaceae. From Gk. *aporos* (impenetrable) and *Cactus.*
   flagelliformis, fla-jel-lee-*form*-is.
   Whip-like (slender stems). Rat's-tail
   Cactus.

Aptenia, ap-*teen*-ee-a. *Aizoaceae.*
From Gk. *apten* (wingless), the capsules have no wings. Tender perennial
succulent.
   cordifolia, kor-di-*fo*-lee-a. Heart-
   shaped (leaves).

Aquilegia, a-kwi-*lee*-jee-a.
Ranunculaceae. From L. *aquila*
(eagle), after the petal shape. Perennial
herbs. Columbine.
   alpina, al-*pie*-na. Alpine.
   canadensis, kan-a-*den*-sis. Of
   Canada. Honeysuckle.
   chrysantha, kris-*anth*-a. With golden
   flowers.
   flabellata, fla-bel-*la*-ta. Fan-shaped.
   longissima, long-*is*-si-ma. Longest.
   scopulorum, skop-ew-*lo*-rum.
   Growing on cliffs.
   vulgaris, vul-*ga*-ris. Common.
   Columbine, Granny's Bonnets.

Aquilegia vulgaris

Arabis, *a*-ra-bis. *Cruciferae*. Origin unknown. Evergreen perennial herbs. Rockcress.
> *albida,* *al*-bi-da. White.
> *blepharophylla,* ble-fa-ro-*fil*-a. With fringed leaves.
> *caucasica,* kaw-*kas*-i-ka. Of the Caucasus.
> *ferdinandii-coburgii,* fer-di-*nan*-dee-ee-ko-*burg*-ee-ee. After King Ferdinand of Bulgaria.

Aralia, a-*ray*-lee-a. *Araliaceae*. Origin unknown. Deciduous trees, shrubs and perennial herbs.
> *elata,* e-*la*-ta. Tall. Japanese Angelica Tree.
> *elegantissima,* e-le-gan-*tis*-i-ma. Most elegant.
> *japonica,* ja-*pon*-i-ka. Of Japan.
> *sieboldii,* see-*bold*-ee-ee. After Siebold.

Araucaria, a-raw-*ka*-ree-a. *Araucariaceae*. After the Chilean Araucani Indians. Semi-hardy conifers.
> *araucana,* a-raw-*ka*-na. After the Araucani. Monkey Puzzle.
> *heterophylla,* he-te-ro-*fil*-a. Variably-leaved. Norfolk Island Pine.

Araujia, a-*raw*-jee-a. *Asclepiadaceae*. The Brazilian name. Evergreen climber.
> *sericifera,* se-ri-*si*-fe-ra. Silk-bearing (hairs on the shoots). Cruel Plant.

Arbutus, *ar*-bu-tus. *Ericaceae*. The L. name. Evergreen trees and shrubs.
> *andrachne,* an-*drak*-nee. Gk. name.
> x *andrachnoides,* an-drak-*noi*-deez. Like *A. andrachne*.
> *menziesii,* men-*zeez*-ee-ee. After Menzies. Madrone.
> *unedo,* *ew*-nee-do. L. name.

Strawberry Tree.

Arctostaphylos, ark-to-*sta*-fil-os. *Ericaceae*. From Gk. *arctos* (bear) and *staphyle* (bunch of grapes). Evergreen trees and shrubs.
> *alpina,* al-*pie*-na. Alpine. Black Bearberry.
> *manzanita,* man-za-*neet*-a. The native Spanish name.
> *nevadensis,* nev-a-*den*-sis. Of the Sierra Nevada, California.
> *patula,* pat-*ew*-la. Spreading.
> *uva-ursi,* *oo*-va-*ur*-see. Bear's grape. Common Bearberry.

Arctostaphylos uva-ursi

Arctotis, ark-*to*-tis. *Compositae*. From Gk. *arctos* (bear) and *otus* (ear). Annual and perennial herbs. African Daisy,
> *breviscapa,* brev-ee-*ska*-pa. With a short scape.
> *stoechadifolia,* stee-ka-di-*fo*-lee-a. With leaves like *Lavandula stoechas*.

Arenaria, a-ree-*na*-ree-a. *Caryophyllaceae*. From L. *arena* (sand), some species grow in sandy areas. Low growing perennial herbs. Sandwort.

balearica, ba-lee-*a*-ri-ka. Of the Balearic Islands.

montana, mon-*ta*-na. Of mountains.

purpurascens, pur-pur-*ras*-ens. Purplish (flowers).

tetraquetra, tet-ra-*kwee*-tra. With leaves in fours.

Argemone, ar-ge-*mo*-nee. *Papaveraceae.* From Gk. *argena* (cataract), an alleged cure. Annual herbs.

mexicana, mex-i-*ka*-na. Mexican. Mexican Poppy.

Arisaema, a-ris-*ee*-ma. *Araceae.* From Gk. *aron* (arum) and *haema* (blood). Tuberous, perennial herbs.

candidissimum, kan-di-*dis*-si-mum. Most white (spathe).

ringens, rin-jens. Gaping (spathe).

triphyllum, tri-*fil*-lum. With three leaves. Indian Turnip.

Arisarum, a-*ris*-a-rum. *Araceae.* From *arisaron* Gk. name for *A. vulgare.* Tuberous perennial herb.

proboscideum, pro-bo-*ski*-dee-um. Like an elephant's trunk (spadix). Mouse Plant.

vulgare, vul-*ga*-ree. Common. Friar's Cowl.

Aristolochia, a-ris-to-*lok*-ee-a. *Aristolochiaceae.* From Gk. *aristos* (best) and *lochia* (childbirth), after its alleged healing properties. Hardy and tender evergreen or deciduous climbers. Birthwort.

elegans, e-le-ganz. Elegant. Calico Flower.

grandiflora, gran-di-*flo*-ra. With large flowers. Pelican Flower.

Armeria, ar-*me*-ree-a. *Plumbaginaceae.* L. name for a *Dianthus.* Evergreen perennial herbs and sub-shrubs.

juniperifolia, joo-ni-pe-ri-*fo*-lee-a. *Juniperus*-leaved.

maritima, ma-*ri*-ti-ma. Growing near the sea. Thrift, Sea Pink.

pseudarmeria, sood-ar-*me*-ree-a. False *Armeria.*

Armeria maritima

Arnebia, ar-*nee*-bee-a. *Boraginaceae.* From the Arabian name. Perennial herbs.

pulchra, pul-kra. Pretty. Prophet Flower.

Arnica, *ar*-ni-ka *Compositae.* From Gk. *arnakis* (lambskin), after the soft leaf texture. Rhizomatous perennial herbs.

montana, mon-*ta*-na. Of mountains.

Aronia, a-*ro*-nee-a *Rosaceae.* From Gk. *aria* (Whitebeam). Deciduous shrubs. Chokeberry.

arbutifolia, ar-bew-ti-*fo*-lee-a. *Arbutus*-leaved. Red chokeberry.

melanocarpa, me-la-no-*kar*-pa. With

black fruits. Black Chokeberry.
prunifolia, proon-i-*fo*-lee-a. Prunus-
leaved. Purple-fruited Chokeberry.

Arrhenatherum, a-ren-*a*-the-rum
Gramineae. From Gk. *arren* (male)
and *ather* (bristle). Perennial grass.
  elatius, e-*la*-tee-us. Tall. False Oat
  Grass.

Arrhenatherum elatius

Artemisia, ar-te-*mis*-ee-a.
Compositae. After Artemis, the Gk.
goddess. Perennial herbs, shrubs and
sub-shrubs.
  abrotanum, a-*brot*-a-num. The L.
  name. Southernwood.
  absinthium, ab-*sin*-thee-um. The L.
  name. Absinthe, Common
  Wormwood.
  arborescens, ar-bo-*res*-enz.
  Becoming tree-like.
  canescens, ka-*nes*-enz. Greyish-
  white hairs.
  frigida, fri-ji-da. Of cold regions.
  lactiflora, lak-ti-*flo*-ra. Milk-flow-
  ered. White Mugwort.
  ludoviciana, loo-do-vik-ee-*a*-na. Of
  Louisiana. White Sage.

schmidtiana, shmit-ee-*a*-na. After
Schmidt.
stelleriana, stel-la-ree-*a*-na. After
Georg Wilhelm Steller. Beach
Wormwood.

Arthropodium, arth-ro-*po*-dee-um.
Liliaceae. From Gk. *arthron* (joint)
and *podion* (stalk), the jointed
pedicels. Perennial herb.
  cirrhatum, si-*ra*-tum. With tendrils.
  Rock Lily.

Arum, *a*-rum. *Araceae.* From Gk.
aron. Tuberous perennial herbs.
  alpinum, al-*pie*-num. Alpine.
  creticum, kree-ti-kum. Of Crete.
  italicum, i-*ta*-li-kum. Of Italy.
  maculatum, mak-ew-*la*-tum. Spotted
  (spathe). Cuckoo Pint, Lords and
  Ladies.
  pictum, pik-tum. Painted (leaves).

Arum maculatum

Aruncus, a-*run*-kus. *Rosaceae.* The
Gk. name. Perennial herbs. Goat's
Beard.

dioicus, dee-o-*ee*-kus. Dioecious.
sinensis, sin-*en*-sis. Of China.

Arundinaria, a-run-di-*na*-ree-a.
Gramineae. From L. *arundo* (reed).
Bamboos.
  anceps, an-seps. Two-headed.
  japonica, ja-*pon*-i-ka. Of Japan.
  nitida, ni-ti-da. Shining (leaves).
  variegata, va-ree-a-*ga*-ta. Variegated
  (leaves).
  viridistriata, vi-ri-dee-stri-*a*-ta.
  Green-striped (leaves).

Arundo, a-*run*-do. *Gramineae.* From
L. *arundo* (reed). Semi-hardy perennial grass.
  donax, do-nax. Gk. name for a reed.

Asarina, a-*sa*-ri-na. *Scrophulariaceae.*
From Spanish name for an *Antirrhinum.*
Evergreen climbers and perennial
herbs. Twining Snapdragon.
  barclaiana, bark-lay-*a*-na. After
  Robert Barclay
  erubescens, e-roo-*bes*-enz. Blushing.
  Creeping Gloxinia.

Asarum europaeum

procumbens, pro-*kum*-benz. Prostrate.
scandens, skan-denz. Climbing.

Asarum, a-*sa*-rum. *Aristolochiaceae.*
From Gk. *asaron.* Rhizomatous perennial herbs. Wild Ginger.
  canadense, kan-a-*den*-see. Of
  Canada. Wild Ginger.
  caudatum, kaw-*da*-tum. With a tail.
  europaeum, ew-ro-*pee*-um. Of
  Europe. Asarabacca.
  hartwegii, hart-*weg*-ee-ee. After Karl
  Theodore Hartweg.
  maximum, max-i-mum. Largest.
  virginicum, vir-*jin*-i-kum. Of
  Virginia.

Asclepias, a-*sklee*-pee-as.
Asclepiadaceae. From Gk. *Asklepios,*
god of medicine. Tender perennial
herbs and sub-shrubs. Silk Weed,
Milkweed.
  curassavica, ku-ra-*sa*-vi-ka. Of
  Curacao. Blood Flower.
  fruticosa, froo-ti-*ko*-sa. Shrubby.
  incarnata, in-kar-*na*-ta. Flesh pink
  (flowers). Swamp Milkweed.
  purpurascens, pur-pur-*ras*-enz.
  Purplish. Purple Silkweed.
  rubra, rub-ra. Red.
  tuberosa, tew-be-*ro*-sa. Tuberous
  (root). Butterfly Weed.

Asimina, a-*si*-mi-na. *Annonaceae.*
From the Native American name
assimin. Deciduous or evergreen trees
and shrubs.
  triloba, tri-*lo*-ba. Three-lobed
  (calyx).

Asparagus, a-*spa*-ra-gus. *Liliaceae.*
The L. name. Hardy and tender perennial herbs and climbers.
  albus, al-bus. White.
  densiflorus, dens-i-*flo*-rus. Densely-
  flowered.

Asparagus officinalis

sprengeri, spreng-a-ree. After Carl
L. Sprenger.
officinalis, o-fi-si-*na*-lis. Sold in
shops. Garden Asparagus.
racemosus, ra-see-*mo*-sus. With
flowers in racemes.
scandens, skan-denz. Climbing.

Asperula, a-*spe*-ru-la. *Rubiaceae.*
From L. *asper* (rough), the rough
stems. Annual and perennial herbs.
hexaphylla, hex-a-*fil*-a. Six-leaved
(whorls of six).
odorata, o-do-*ra*-ta. Scented.
suberosa, soo-be-*ro*-sa. Corky-
stemmed.

Asphodeline, as-fod-e-*lee*-nee.
Liliaceae. Perennial herbs. Jacob's Rod.
liburnica , li-*burn*-i-ka.From
Liburnia (Croatia).
lutea, loo-tee-a. Yellow (flowers).
Yellow Asphodel.

Aspidistra, a-spi-*di*-stra. *Liliaceae.*
From Gk. *aspideon* (a small, round
shield), after the stigma shape. Tender,

rhizomatous evergreen herb.
elatior, e-*la*-tee-or. Taller.

Asplenium, a-*splee*-nee-um.
Aspleniaceae. From Gk. *a* (not) and
splen (spleen), after its alleged healing
properties. Hardy and tender evergreen and semi-evergreen ferns.
Spleenwort.
bulbiferum, bul-*bi*-fe-rum. Producing
bulbs. Mother Spleenwort.
marinum, ma-*reen*-num. Growing
near the sea. Sea Spleenwort.
nidus, nee-dus. A nest. Bird's Nest
Fern.
trichomanes, tri-ko-*ma*-neez. Gk.
name for a fern. Maidenhair
Spleenwort.

Aster, *a*-ster. *Compositae.* From L.
aster (star), after the flower shape.
Perennial herbs and deciduous or evergreen sub-shrubs. Michaelmas Daisy.
acris, a-kris. Sharp-tasting.
albescens, al-*bes*-enz. Whitish
(under the leaves).
amellus, a-*mel*-lus. The L. name.
cordifolius, kor-di-*fo*-lee-us. With

Asperula odorata

heart-shaped leaves.
ericoides, e-ri-*koi*-deez. *Erica*-like
frikartii, fri-*kart*-ee-ee. After Carl
Ludwig Frikart.
novi-belgii, no-vee-*bel*-jee-ee. Of
New York. Michaelmas Daisy.
paniculatus, pa-nik-ew-*la*-tus. With
flowers in panicles.
sedifolius, se-di-*fo*-lee-us. *Sedum*-
leaved.
thomsonii, tom-*son*-ee-ee. After
Thomas Thomson.

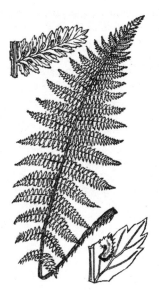

Athyrium filix-femina

Astilbe, a-*stil*-bee. *Saxifragaceae.*
From Gk. *a* (without) and *stilbe* (bril-
liance), after the dull leaves. Perennial
herbs.
chinensis, chin-*en*-sis. Of China.
simplicifolia, sim-pli-ki-*fo*-lee-a.
With simple leaves.

Astrantia, a-*stran*-tee-a. *Umbelliferae.*
Origin unknown. Perennial herbs.
Masterwort.

major, ma-jor. Larger. Greater
Masterwort.
maxima, max-i-ma. Largest.

Astrophytum, a-*stro*-fi-tum.
Cactaceae. From Gk. *astron* (star) and
phyton (plant), after the star-shaped
body.
ornatum, or-*na*-tum. Ornamental.

Athyrium, a-*thi*-ree-um. *Woodsiaceae.*
Origin unknown. Lady Fern.
filix-femina, fi-lix-*fe*-mi-na. Lady
fern. Lady Fern.
goeringianum, gur-ring-gee-*a*-num.
After Goering.
nipponicum, ni-*pon*-i-kum. Of Japan.
pictum, *pik*-tum. Painted (leaves).
Japanese Painted Fern.

Atriplex, *a*-tri-plex. *Chenopodiaceae.*
Gk. name. Annual or perennial herbs
and evergreen or semi-evergreen shrubs.
canescens, ka-*nes*-enz. Greyish-
white hairs.
halimus, ha-li-mus. The Gk. name.
Tree Purslane.
hortensis, hor-*ten*-sis. Of gardens.
Mountain Spinach.

Atriplex hortensis

Aubrieta, aw-bree-*she*-a. *Cruciferae.*
After Claude Aubriet. Evergreen, trailing, perennial herbs.
  deltoidea, del-*toi*-dee-a. Triangular
  (petals).
  gracilis, gra-si-lis. Graceful.

Aucuba, aw-*kew*-ba. *Cornaceae.* From
the Japanese name. Evergreen shrub.
  japonica, ja-*pon*-i-ka. Of Japan.

Aurinia, aw-*rin*-ee-a. *Cruciferae.*
From L. *aureus* (golden), the flowers.
Evergreen perennial herbs.
  saxatilis, sax-*a*-ti-lis. Growing
  among rocks.

Austrocedrus, aw-stro-*sed*-rus.
Cupressaceae. From L. *australis* (southern) and *Cedrus.* Evergreen conifer.
  chilensis, chil-*en*-sis. Of Chile.
  Chilean Cedar.

Azara, a-*za*-ra. *Flacourtiaceae.* After
J. N. Azara. Evergreen, semi-hardy
trees and shrubs.
  lanceolata, lan-see-o-*la*-ta. Spear-
  shaped.
  microphylla, mie-kro-*fil*-a. Small-
  leaved.
  serrata, se-*ra*-ta. Saw-toothed.

Azolla, a-*zo*-la. *Azollaceae.* From Gk.
azo (dry) and *ollua* (kill), they die
when dry. Deciduous, perennial, floating, aquatic ferns.
  caroliniana, ka-ro-lin-ee-*a*-na. Of
  Carolina.

B

Babiana, bab-ee-*a*-na. *Iridaceae.*
From Afrikaans *babiaans* (baboon),
said to eat the corms. Semi-hardy, cormous herbs. Baboon Flower.
    plicata, pli-*ka*-ta. Pleated (leaves).
    stricta, strik-ta. Erect (stems).

Baccharis, *ba*-ka-ris. *Compositae.*
After Bacchus, god of wine.
Deciduous or evergreen shrubs.
    halimifolia, ha-li-mi-*fo*-lee-a. With
leaves like *Atriplex halimus.*
Groundsel Tree.

Ballota, ba-*lo*-ta. *Labiatae.* Gk. name
for *B. nigra.* Perennial herbs and
deciduous or evergreen sub-shrubs.
    pseudodictamnus, soo-do-dik-*tam*-
nus. False *Dictamnus.*
    nigra, *nig*-ra. Black.

Banksia, *bank*-see-a. *Proteaceae.*
After Sir Joseph Banks. Australian
Honeysuckle.
    coccinea, kok-*kin*-ee-a. Scarlet.
Scarlet Banksia.
    repens, *ree*-penz. Creeping.
    serrata, se-*ra*-ta. Saw-toothed
(leaves). Saw Banksia.

Baptisia, bap-*tis*-ee-a. *Leguminosae.*
From Gk. *bapto* (dye). Perennial
herbs.
    australis, aw-*stra*-lis. Southern. Blue
False Indigo.
    tinctoria, tink-*to*-ree-a. Used in dye-
ing. Wild Indigo.

Bauera, *baw*-a-ra. *Cunoniaceae.* After
the brothers Franz and Ferdinand
Bauer. Tender, evergreen shrubs.
    rubioides, roo-bee-*oi*-deez. *Rubia*-like.

Bauhinia, baw-*hin*-ee-a. *Leguminosae.*
After John and Caspar Bauhin. Tender
trees, shrubs and climbers.
    blakeana, blayk-ee-*a*-na. After Sir
Henry and Lady Blake.
    galpinii, gal-*pin*-ee-ee. After Ernst
E. Galpin
    punctata, punk-*ta*-ta. Spotted.
    purpurea, pur-*pur*-ree-a. Purple
(flowers). Butterfly Tree.
    variegata, va-ree-a-*ga*-ta. Variegated
(flowers). Orchid Tree.

Beaumontia, bo-*mont*-ee-a.
Apocynaceae. After Lady Diana
Beaumont. Tender, evergreen climbers.
    grandiflora, gran-di-*flo*-ra. Large-
flowered. Nepal Trumpet Flower.

Begonia, bee-*gon*-ee-a. *Begoniaceae.*
After Michael Begon. Tender perennial
herbs and shrubs.
    angularis, ang-gew-la-ris. Angular.
    boliviensis, bo-liv-ee-*en*-sis. Of
Bolivia.
    boweri, *bow*-a-ree. After Bower.
Eyelash Begonia.
    coccinea, kok-*kin*-ee-a. Scarlet.
Angel-wing Begonia.
    corallina, ko-ra-*leen*-a. Coral-red.
    davisii, day-*vis*-ee-ee. After Walter
Davis.
    diadema, die-a-*dee*-ma. A crown.
    dregei, *dree*-gee-ee. After Johann
Franz Drege. Maple-leaf Begonia.
    x *erythrophylla,* e-rith-ro-*fil*-a. Red-
leaved. Kidney Begonia.
    foliosa, fo-lee-*o*-sa. Leafy.
    fuchsioides, few-she-*oi*-deez.
Fuchsia-like. Fuchsia Begonia.
    haageana, harg-ee-*a*-na. After Haage.
    maculata, mak-ew-*la*-ta. Spotted.

manicata, man-i-*ka*-ta. Long-sleeved.

masoniana, may-son-ee-*a*-na. After L. Maurice Mason. Iron Cross Begonia.

metallica, me-*ta*-li-ka. Metallic (leaves). Metallic-leaf Begonia.

rex, rex. Of the King.

Rex-cultorum, rex-kul-*to*-rum. Cultivated *B. rex.*

scharffii, sharf-ee-ee. After Carl Scharff. Elephant's-ear Begonia.

semperflorens, sem-per-*flo*-renz. Ever-flowering.

serratipetala, se-ra-tee-*pe*-ta-la. Toothed petals.

socotrana, so-ko-*tra*-na. Of Socotra.

zebrina, ze-*breen*-a. Striped.

Belamcanda, bel-am-*kan*-da. *Iridaceae.* From the Asian name. Perennial herbs.

chinensis, chin-*en*-sis. Of China. Leopard Lily.

Bellis perennis

Bellis, *bel*-is. *Compositae.* From L. *bellus* (pretty). Perennial herbs.

perennis, pe-*re*-nis. Perennial. Daisy.

Bellium, *bel*-ee-um. *Compositae.* From *Bellis.* Annual or perennial herbs.

bellidioides, bel-i-dee-*oi*-deez. The False Daisy.

minutum, mi-*new*-tum. Small.

Berberidopsis, ber-be-ri-*dop*-sis. *Flacourtiaceae.* From *Berberis* and Gk. *-opsis* (resemblance). Evergreen climbers.

corallina, ko-ra-*lee*-na. Coral-red. Coral Plant.

Berberis, *ber*-be-ris. *Berberidaceae.* From the Arabic. Deciduous, ever-green and semi-evergreen shrubs.

aggregata, ag-re-*ga*-ta. Clustered.

buxifolia, bux-i-*fo*-lee-a. *Buxus*-leaved.

darwinii, dar-*win*-ee-ee. After Charles Darwin.

gagnepainii, gan-ya-*pan*-ee-ee. After Francois Gagnepain.

hookeri, huk-a-ree. After W. J. Hooker.

ilicifolia, i-lis-i-*fo*-lee-a. Holly-leaved.

julianae, joo-lee-*a*-nee. After Juliana Schneider.

linearifolia, lin-ee-a-ri-*fo*-lee-a. With narrow leaves.

Berberis vulgaris

ottawensis, o-ta-*wen*-sis. Of Ottawa.
sargentiana, sar-jen-tee-*a*-na. After
Sargent.
stenophylla, sten-o-*fil*-a. Narrow-
leaved.
thunbergii, thun-*berg*-ee-ee. After
Thunberg.
verruculosa, ve-roo-kew-*lo*-sa. With
small warts (on the shoots).
vulgaris, vul-*ga*-ris. Common.
Common Barberry.
wilsoniae, wil-*so*-nee-ee. After Mrs
Ernest H. Wilson.

Bergenia, ber-*gen*-ee-a.
Saxifragaceae. After Karl August von
Bergen. Evergreen perennial herbs.
ciliata, si-lee-*a*-ta. Fringed with
hairs (leaves).
crassifolia, kra-si-*fo*-lee-a. Thick-
leaved.
purpurascens, pur-pur-*ras*-enz.
Purplish.
x *schmidtii, shmit*-ee-ee. After Ernst
Schmidt.
stracheyi, stray-kee-ee. After
Lieutenant-General Sir Richard
Strachey.

Bertolonia, ber-to-*lo*-nee-a.
Melastomataceae. After Antonio
Bertoloni. Tender, evergreen perennial
herbs.
maculata, mak-ew-*la*-ta. Spotted
(leaves).
marmorata, mar-mo-*ra*-ta. Marbled
(leaves).

Bessera, *bes*-a-ra. *Liliaceae.* After
Wilibald von Besser. Tender cormous
herbs.
elegans, e-le-ganz. Elegant. Coral
Drops.

Beta, *bee*-ta. *Chenopodiaceae.* L.
name. Biennial or perennial herbs.

vulgaris, vul-*ga*-ris. Common.
Beetroot.

Betula, *bet*-ew-la. *Betulaceae.* L.
name. Deciduous trees and shrubs.
Birch.
albo-sinensis, al-bo-si-*nen*-sis.
Chinese *B. alba.* Chinese Red Birch.
cordifolia, kor-di-*fo*-lee-a. With
heart-shaped leaves.
ermanii, er-*man*-ee-ee. After Adolph
Erman. Gold Birch.
lutea, loo-tee-a. Yellow (bark).
Yellow Birch.
nana, na-na. Dwarf. Dwarf Birch.
nigra, ni-gra. Black (bark). Black
Birch.
papyrifera, pa-pi-*ri*-fe-ra. Paper-
bearing. White Birch, Paper Birch.
pendula, pen-dew-la. Pendulous.
Silver Birch.
utilis, ew-ti-lis. Useful.

Betula pendula

Bifrenaria, bi-free-*na*-ree-a.
Orchidaceae. From L. *bis* (twice)
frenum (strap). Evergreen, epiphytic

greenhouse orchids.
harrisoniae, ha-ri-*so*-nee-ee. After
Mrs Arnold Harrison.

Bignonia, big-*no*-nee-a. *Bignoniaceae.*
After Abbé Jean Paul Bignon.
Evergreen climbers.
capreolata, ka-pree-o-*la*-ta. Bearing
tendrils. Trumpet Flower.

Billardiera, bi-lar-dee-*e*-ra.
Pittosporaceae. After J. J. H. de
Labillardière. Semi-hardy, evergreen
climbers.
longiflora, long-i-*flo*-ra. Long-flow-
ered. Blueberry.

Billbergia, bil-*berg*-ee-a.
Bromeliaceae. After J. G. Billberg.
Tender, epiphytic, perennial herbs.
elegans, e-le-ganz. Elegant.
nutans, new-tanz. Nodding.
Friendship Plant.
vittata, vi-*ta*-ta. Banded (leaves).
x *windii, vin*-dee-ee. After Wind.
zebrina, ze-*bree*-na. Zebra-striped.

Blechnum spicant

Blechnum, *blek*-num. *Blechnaceae.*
From *blechnon* Gk. name for a fern.
Hardy and tender, evergreen or semi-
evergreen ferns. Hard Fern.
brasiliense, bra-zil-ee-*en*-see. Of
Brazil.
occidentale, ok-si-den-*ta*-lee.
Western. Hammock Fern.
magellanica, ma-jel-*an*-i-ka. From
the region of the Magellan Straits.
spicant, spi-kant. Tufted. Deer Fern.

Boenninghausenia, burn-ing-how-
zen-ee-a. *Rutaceae.* After von
Boenninghausen. Deciduous sub-
shrub.
albiflora, al-bi-*flo*-ra. White-flow-
ered.

Bomarea, bo-*ma*-ree-a.
Alstroemeriaceae. After Valmont de
Bomare. Tuberous, perennial herbs
and climbers.
caldasii, kal-*da*-see-ee. After
Francisco Jose de Caldas.

Borago, bo-*ra*-go, *Boraginaceae.*
From L. *burra* (rough hair). Annual
and perennial herbs.
laxiflora, lax-i-*flo*-ra. Loose-flow-
ered.
officinalis, o-fi-si-*na*-lis. Sold in
shops. Borage.

Boronia, bo-*ro*-nee-a. *Rutaceae.* After
Francesca Borone. Tender, evergreen
shrubs.
crenulata, kren-ew-*la*-ta. Scalloped.
elatior, e-*la*-tee-or. Taller.
heterophylla, he-te-ro-*fil*-a. With
variable-shaped leaves. Red
Boronia.
megastigma, meg-a-*stig*-ma. With
large stigma. Scented Boronia.
serrulata, se-ru-*la*-ta. Small-toothed
(leaves). Sydney Rock Rose.

Borago officinalis

Borzicactus, bor-zee-*kak*-tus.
Cactaceae. After Antonio Borzi and
Cactus.
　aurantiacus, aw-ran-tee-*a*-kus.
　Orange.
　haynei, hayn-ee-ee. After Frederich
　Hayne.

Bougainvillea, boo-gan-*vil*-lee-a.
Nyctaginaceae. After Louis Antoine de
Bougainville. Tender, deciduous or
evergreen climbers.
　x *buttiana,* but-ee-*a*-na. After Mrs R.
　V. Butt
　glabra, gla-bra. Smooth.
　spectabilis, spek-*ta*-bi-lis.
　Spectacular.

Bouvardia, boo-*var*-dee-a. *Rubiaceae*.
After Dr Charles Bouvard. Tender,
deciduous semi-evergreen and evergreen shrubs.
　longiflora, long-i-*flo*-ra. Long-flow-
　ered. Sweet Bouvardia.
　ternifolia, tern-i-*fo*-lee-a. With
　leaves in threes. Scarlet Trompetilla.
　triphylla, tri-*fil*-a. With three leaves.

Brachycome, bra-kee-*ko*-mee.
Compositae. From Gk. *brachys* (short)
and *kome* (hair). Annual and perennial
herbs.
　iberidifolia, i-be-ri-di-*fo*-lee-a.
　With *Iberis*-like leaves. Swan River
　Daisy.

Brachyglottis, bra-kee-*glo*-tis.
Compositae. From Gk. *brachys* (short)
and *glotta* (tongue) after the short ray
florets. Semi-hardy, evergreen trees
and shrubs.
　repanda, re-*pan*-da. With wavy mar-
　gins (leaves).

Brassavola, bra-*sa*-vo-la.
Orchidaceae. After Antonio
Brassavola, Greenhouse orchids.
　digbyana, dig-bee-*a*-na. After
　Edward Digby.
　nodosa, no-*do*-sa. Conspicuous
　nodes. Lady of the Night.

Brassia, *brass*-ee-a. *Orchidaceae*.
After William Brass. Greenhouse
orchids.
　maculata, mak-ew-*law*-ta. Spotted
　(petals).
　verrucosa, ve-roo-*ko*-sa. Warty.

Brassica, *bra*-si-ka. *Cruciferae*. L.
name for cabbage. Annual, biennial
and perennial herbs.
　acephala, a-*sef*-a-la. Without a head.
　Kale.
　botrytis, bot-ri-tis. Like a bunch of
　grapes. Broccoli, Cauliflower.
　capitata, ka-pi-*ta*-ta. In a dense
　head. Cabbage.
　gemmifera, jem-*i*-fe-ra. Bearing
　buds. Brussels Sprouts.
　gongylodes, gon-gi-*lo*-deez.
　Swollen. Kohl Rabi.
　italica, ee-*ta*-li-ka. Of Italy.
　Sprouting Broccoli.

oleracea, o-le-*ra*-see-a. Vegetable-like. Wild Cabbage.
pekinensis, pee-kin-*en*-sis. Of Peking. Chinese Cabbage.
perviridis, per-*vi*-ri-dis. Very green. Tendergreen.
rapa, ra-pa. L. name. Turnip.

Brimeura, bri-*mur*-ra. *Liliaceae.* After Maria de Brimeur. Bulbous perennial.
   amethystina, a-me-*this*-ti-na. Violet.

Briza, *bree*-za. *Gramineae.* Gk. name. Annual and perennial grasses.
   maxima, max-i-ma. Largest. Greater Quaking Grass.
   media, me-dee-a. Intermediate. Common Quaking Grass.
   minor, mi-nor. Smaller. Lesser Quaking Grass.

Briza media

Brodiaea, bro-dee-*ee*-a. *Liliaceae.* After James Brodie. Cormous perennial herbs.
   coronaria, ko-ro-*na*-ree-a. Of garlands.

ida-maia, ee-da-*ma*-ya. After Ida May Burke.
pulchella, pul-*kel*-la. Pretty.

Browallia, bro-*a*-lee-a. *Solonaceae.* After Bishop John Browall. Tender perennials.
   elata, e-*la*-ta. Tall.
   speciosa, spes-ee-*o*-sa. Showy. Sapphire Flower.

Bruckenthalia, bruk-an-*thal*-ee-a. *Ericaceae.* After Samuel and Michael von Bruckenthal. Evergreen, shrub.
   spiculifolia, spik-ew-lee-*fo*-lee-a. Spiky-leaved. Spike Heath.

Brunfelsia, brun-*fel*-see-a. *Solanaceae.* After Otto Brunfels. Tender, evergreen trees or shrubs.
   calycina, ka-li-*see*-na. With a well-developed calyx.
   pauciflora, paw-si-*flo*-ra. With few flowers.

Brunnera, *brun*-er-a. *Boraginaceae.* After Samuel Brunner. Perennial herbs.
   macrophylla, mak-ro-*fil*-a. Large-leaved.

Buddleia, *bud*-lee-a. *Loganiaceae.* After the Rev. Adam Buddle. Deciduous, semi-evergreen or evergreen trees and shrubs. Butterfly Bush.
   alternifolia, al-tern-i-*fo*-lee-a. With alternate leaves on each side of the stem.
   colvilei, kol-*vil*-ee-ee. After Sir James Colvile.
   crispa, kris-pa. Finely waved (leaves).
   davidii, da-*vid*-ee-ee. After Armand David. Butterfly Bush.
   fallowiana, fa-lo-ee-*a*-na. After George Fallow.

globosa, glo-*bo*-sa. Spherical. Orange Ball Tree.

x *weyeriana,* way-a-ree-*a*-na. After Van de Weyer.

Bulbophyllum, bul-bo-*fil*-lum. *Orchidaceae.* From Gk. *bolbos* (bulb) and *phyllon* (leaf). The leaves grow from a pseudobulb. Greenhouse orchids.

   careyanum, kair-ree-*a*-num. After Dr Carey.

Buphthalmum, buf-*thal*-mum. *Compositae.* From Gk. *bous* (ox) and *ophthalmos* (eye). Perennial herbs. Ox Eye.

   salicifolium, sa-li-si-*fo*-lee-um. *Salix*-leaved.

   speciosum, spes-ee-*o*-sum. Showy.

Bupleurum, boo-*plur*-rum. *Umbelliferae.* From Gk. *boupleuros* (ox-rib). Evergreen shrub. Thorow Wax.

   fruticosum, froo-ti-*ko*-sum. Shrubby. Shrubby Hare's Ear.

Butia, *bew*-tee-a. *Palmae.* Origin unknown. Tender, evergreen palms. Jelly Palm.

   capitata, ka-pi-*ta*-ta. A dense head.

Butomus, *boo*-to-mus. *Butomaceae.* From Gk. *bous* (ox) and *temmo* (cut).

Perennial aquatic herb. Water Gladiolus.

   umbellatus, um-bel-*a*-tus. Flowers in umbels.

Buxus sempervirens

Buxus, *bux*-us. *Buxaceae.* L. name. Evergreen trees and shrubs. Box.

   balearica, ba-lee-*a*-ri-ka. Of the Balearic Islands. Balearic Boxwood.

   microphylla, mie-kro-*fil*-a. Small-leaved.

   sempervirens, sem-per-*vi*-renz. Evergreen. Common Box.

   wallichiana, wo-lik-ee-*a*-na. After Nathaniel Wallich.

C

Cabomba, ka-*bom*-ba.
Nymphaeaceae. From the native
Guiana name. Deciduous or semi-evergreen aquatic herbs. Fanwort.
  aquatica, a-*kwa*-ti-ka. Growing in
  water.
  caroliniana, ka-ro-lin-ee-*a*-na. Of
  Carolina.

Caesalpinia, see-zal-*pee*-nee-a.
Leguminosae. After Andreas
Caesalpini. Tender trees.
  gilliesii, gi-*leez*-ee-ee. After John
  Gillies.
  pulcherrima, pul-*ke*-ri-ma. Very
  Pretty. Barbados Pride.

Caladium, ka-*la*-dee-um, *Araceae.*
From the Indian *kaladi.* Tender, perennial herbs. Angel's Wings, Elephant's
Ears.
  x *hortulanum,* hort-ew-*la*-num. Of
  gardens.
  x *candidum, kan*-di-dum. White
  (leaves).

Calamintha, kal-a-*min*-tha. *Labiatae.*
From Gk. *kalos* (beautiful) and *minthe*
(mint). Aromatic, perennial herbs.
Calamint.
  alpina, al-*pie*-na. Alpine.
  grandiflora, gran-di-*flo*-ra. Large-
  flowered.

Calanthe, ka-*lan*-thee. *Orchidaceae.*
From Gk. *kalos* (beautiful) and *anthos*
(flower). Greenhouse orchids.
  alpina, al-*pie*-na. Alpine.
  rosea, ro-see-a. Rose-coloured.
  vestita, ves-*tee*-ta. Clothed (hairy
  stem).

Calathea, ka-*la*-thee-a. *Marantaceae.*
From Gk. *kalathos* (basket), the flower
cluster resembles a basket of flowers.
Tender, evergreen perennial herbs.
  bella, be-la. Pretty.
  lindeniana, lin-den-ee-*a*-na. After J.
  J. Linden.
  makoyana, mak-oy-*a*-na. After Jacob
  Makoy. Peacock Plant.
  ornata, or-*na*-ta. Showy.
  zebrina, zeb-*ree*-na. Striped. Zebra
  Plant.

Calceolaria, kal-see-o-*la*-ree-a.
Scrophulariaceae. After F. Calceolari.
From L. *calceolus* (slipper), after the
flower shape. Annual, biennial and
evergreen perennial herbs and sub-
shrubs. Slipperwort.
  biflora, bi-*flo*-ra. Two-flowered.
  darwinii, dar-*win*-ee-ee. After
  Charles Darwin.
  x *herbeohybrida,* herb-ee-o-*hib*-ri-
  da. Herbaceous hybrid.
  integrifolia, in-teg-ri-*fo*-lee-a. With
  entire leaves.
  tenella, ten-*el*-la. Dainty.

Calendula, kal-*en*-dew-la.
Compositae. From L. *calendae* (first
day of the month), after its long flowering period. Annual herb.
  fruticosa, froo-ti-*ko*-sa. Shrubby.
  officinalis, o-fi-si-*na*-lis. Sold in
  shops. Pot Marigold.

Calla, ka-la. *Araceae.* From Gk. *kalos*
(beautiful). Deciduous or semi-evergreen perennial herbs.
  palustris, pa-*lus*-tris. Growing in
  marshes. Bog Arum.

Callicarpa, ka-lee-*kar*-pa.
Verbenaceae. From Gk. *kalos* (beautiful) and *karpos* (fruit), after their beautiful fruit. Deciduous shrubs. Beauty Berry.
　bodinieri, bo-din-ee-*e*-ree. After Emile Bodinieri.
　japonica, ja-*pon*-i-ka. Of Japan.

Callisia, kal-*is*-ee-a. *Commelinaceae.* From Gk. *kallis* (beauty). Tender, evergreen perennial herbs. Inch Plant.
　elegans, *e*-le-ganz. Elegant. Striped Inch Plant.
　fragrans, *fra*-granz. Fragrant (flowers).

Callistemon, ka-*li*-stee-mon.
Myrtaceae. From Gk. *kalos* (beautiful) and *stemon* (stamen). The flowers are the beautiful part of the shrub. Evergreen, semi-hardy shrubs. Bottlebrush tree.
　citrinus, si-*tri*-nus. Lemon-scented (leaves). Crimson Bottle-brush.
　linearis, lin-ee-*a*-ris. Narrow (leaves).
　pallidus, *pa*-li-dus. Pale.
　rigidus, *ri*-ji-dus. Rigid (leaves).
　speciosus, spes-ee-*o*-sus. Showy.
　subulatus, sub-ew-*la*-tus. Awl-shaped (leaves).

Callistephus, ka-*lee*-ste-fus.
Compositae. From Gk. *kalos* (beautiful) and *stephanus* (crown), after the showy flower heads. Semi-hardy, annual herb.
　chinensis, chin-*en*-sis. Of China. China Aster.

Calluna, ka-*loo*-na. *Ericaceae.* From L. *kalluno* (cleanse), used as a broom. Evergreen shrubs.
　vulgaris, vul-*ga*-ris. Common. Heather, Ling.

Calocedrus, ka-lo-*sed*-rus.
Cupressaceae. From Gk. *kalos* (beautiful) and *Cedrus.* Evergreen conifer.
　decurrens, dee-*ku*-renz. The leaf base merges with the stem.

Calochortus, ka-lo-*kor*-tus. *Liliaceae.* From Gk. *kalos* (beautiful) and *chortos* (grass), after the slender leaves. Semi-hardy bulbous herbs. Mariposa Lily.
　albus, *al*-bus. White.
　amabilis, a-*ma*-bi-lis. Beautiful.
　barbatus, bar-*ba*-tus. Bearded (petals).
　caeruleus, see-*ru*-lee-us. Dark blue.
　luteus, *loo*-tee-us. Yellow.

Caltha, *kal*-tha. *Ranunculaceae.* L. name for a plant with yellow flowers. Deciduous and perennial aquatic herbs.
　asarifolia, a-sa-ri-*fo*-lee-a. With *Asarum*-like leaves.
　chelidonii, kel-i-*do*-nee-ee. *Chelidonium*-like.
　leptosepala, lep-to-*sep*-a-la. With slender sepals.

Calluna vulgaris

41

Caltha palustris

palustris, pa-*lus*-tris. Growing in marshes. King Cup, Marsh Marigold.
polypetala, po-li-*pe*-ta-la. With many petals.

Calycanthus, ka-lee-*kanth*-us. *Calycanthaceae.* From Gk. *kalyx* (calyx) and *anthos* (flower). The sepals and petals are similar in colour. Deciduous shrubs.
fertilis, fer-ti-lis. Fertile.
floridus, flo-ri-dus. Flowering. Carolina Allspice.
occidentalis, ok-si-den-*ta*-lis. Western. Californian Allspice.

Camassia, ka-*mas*-ee-a. *Liliaceae.* From the Native American *quamash.* Bulbous herbs.
cusickii, kew-*sik*-ee-ee. After W. C. Cusick.
leichtlinii, liekt-*lin*-ee-ee. After Max Leichtlin.
quamash, kwa-mash. The Native American name. Common Camassia.
scilloides, sil-*oy*-deez. *Scilla*-like. Wild Hyacinth.

Camellia, ka-*me*-lee-a. *Theaceae.* After George Joseph Kame. Tender to hardy, evergreen trees and shrubs.
cuspidata, kus-pi-*da*-ta. With a stiff point (leaves).
japonica, ja-*pon*-i-ka. Of Japan. Common Camellia.
reticulata, ree-tik-ew-*la*-ta. Net-veined (leaves).
sasanqua, sa-*san*-kwa. The Japanese name.
sinensis, si-*nen*-sis. Of China. Tea Plant.

Campanula, kam-*pan*-ew-la. *Campanulaceae.* From L. *campana* (bell), after the flower shape. Annual, biennial and perennial herbs. Bellflower.
alliariifolia, a-lee-a-ree-i-*fo*-lee-a. With leaves like *Alliaria petiolata.* Spurred Bellflower.
barbata, bar-*ba*-ta. Bearded (corolla).
carpatica, kar-*pa*-ti-ka. Of the Carpathian Mountains.
cochleariifolia, kok-lee-a-ree-i-*fo*-lee-a. With *Cochlearia*-like leaves.
excisa, ex-*see*-sa. Cut away.

Campanula persicifolia

garganica, gar-*ga*-ni-ka. Of Monte Gargano, Italy. Adriatic Bellflower.
glomerata, glo-me-*ra*-ta. Clustered. Clustered Bellflower.
isophylla, i-so-*fil*-a. With equal-sized leaves. Italian Bellflower.
lactiflora, lak-ti-*flo*-ra. Milky-flowered. Milky Bellflower.
latifolia, la-ti-*fo*-lee-a. Broad-leaved. Giant Bellflower.
medium, me-dee-um. Medium-sized. Canterbury Bell.
morettiana, mo-ret-ee-*a*-na. After Moretti.
persicifolia, per-si-ki-*fo*-lee-a. With leaves like *Prunus persica.*
portenschlagiana, por-ten-shlag-ee-*a*-na. After Franz von Portenschlag-Ledermeyer.
poscharskyana, po-shar-skee-*a*-na. After Gustav Adolf Poscharsky.
pulla, pul-la. Dark.
punctata, punk-*ta*-ta. Spotted (corolla).
pyramidalis, pi-ra-mi-*da*-lis. Pyramidal. Chimney Bellflower.
raineri, ray-ne-ree. After Rainer.
rotundifolia, ro-tund-i-*fo*-lee-a. With round leaves. Harebell, Bluebell (Scotland).
sarmatica, sar-*ma*-ti-ka. Of Sarmatia.
trachelium, tra-*ke*-lee-um. From old Gk. name. Throatwort.
zoysii, zoys-ee-ee. After Karl von Zoys.

Campsis, *kamp*-sis. *Bignoniaceae.* From Gk. *kampe* (bent), after the curved stamens. Deciduous climbers.
grandiflora, gran-di-*flo*-ra. Large-flowered.
radicans, ra-di-kanz. With rooting stems. Trumpet Vine.
x *tagliabuana,* tal-ee-a-bew-*a*-na. After the Tagliabue brothers.

Camptosorus, kamp-to-*sor*-us. *Polypodiaceae.* From Gk. *kamptos* (curved) and *sorus,* after the curved sori. Deciduous or semi-evergreen ferns.
rhizophyllus, rie-zo-*fil*-us. Rooting leaves. (The leaves root at the tip and appear to 'walk'). Walking Fern.

Canna, *ka*-na. *Cannaceae.* From Gk. *kanna* (reed). Tender, perennial herbs.
x *generalis,* gen-e-*ra*-lis. Common form.
indica, in-di-ka. Of India. Indian Shot.

Cannabis, *kan*-a-bis. *Cannabaceae.* From Gk. *kannabis* (hemp). Annual herb.
sativa, sa-*tee*-va. Cultivated. Hemp.

Cantua, *kan*-tew-a. *Polemoniaceae.* From the Peruvian name. Semi-hardy evergreen shrubs.
buxifolia, bux-i-*fo*-lee-a. *Buxus*-leaved.

Capsicum, *kap*-si-kum. *Solonaceae.* From Gk. *kapto* (bite), after the spicy taste. Tender, evergreen shrubs.
annuum, an-ew-um. Annual. Sweet Pepper, Christmas Pepper.
frutescens, froo-*tes*-enz. Shrubby. Hot Pepper.

Caragana, ka-ra-*ga*-na. *Leguminosae.* From the Mongolian *Caragan.* Deciduous shrubs.
arborescens, ar-bo-*res*-enz. Tree-like. Pea Tree.
lorbergii, lor-*berg*-ee-ee. After Lorberg's nursery, Germany.

Cardamine, kar-*dam*-i-nee. *Cruciferae.* From the Gk. for a cress plant. Annual and perennial herbs.

Cardamine pratensis

Bitter Cress.
californica, kal-i-*for*-ni-ka. Of
California.
enneaphyllus, en-ee-a-*fil*-us. With
nine leaves.
heptaphylla, hep-ta-*fil*-a. With seven
leaves.
laciniata, la-sin-ee-*a*-ta. Deeply cut
(leaves).
lyrata, li-*ra*-ta. Lyre-shaped (leaves).
pentaphyllus, pen-ta-*fil*-us. With five
leaves.
pratensis, pra-*ten*-sis. Of meadows.
Ladies' Smock, Cuckoo Flower.

Cardiocrinum, kar-dee-o-*kri*-num.
Liliaceae. From Gk. *kardio* (heart) and
krinon (lily), after the heart-shaped
leaves. Bulbous herbs.
cordatum, kor-*da*-tum. Heart-shaped
(leaves).
giganteum, ji-*gan*-tee-um. Very
large.

Carex, *kar*-ex. *Cyperaceae.* The L.
name. Perennial, grass-like herb.
Sedge.
morrowii, mo-*ro*-ee-ee.After Morrow.
pendula, pen-dew-la. Pendulous
(flower spikes).

pseudocyperus, sood-o-sie-*pe*-rus.
False *Cyperus.*
riparia, ree-*pa*-ree-a. Of river banks.
Great Pond Sedge.
sylvatica, sil-*va*-ti-ka. Of woods.

Carex pseudocyperus

Carlina, kar-*lee*-na. *Compositae.* After
Charlemagne. Annual, biennial and
perennial herbs. Carline Thistle.
acanthifolia, a-kanth-i-*fo*-lee-a. With
Acanthus-like leaves.
acaulis, a-*kaw*-lis. Stemless.
vulgaris, vul-*ga*-ris. Common.
Common Carline Thistle.

Carmichaelia, kar-mie-*keel*-ee-a.
Leguminosae. After Capt. Carmichael.
New Zealand shrubs.
australis, aw-*stra*-lis. Southern.
enysii, e-*nis*-ee-ee. After John
Davies Enys.
petriei, pet-ree-ee. After Petrie.

Carnegiea, kar-*nee*-gee-a. *Cactaceae.*
After Andrew Carnegie. Perennial
Cacti.
gigantea, ji-*gan*-tee-a. Very large.

Carpenteria, kar-pen-*ter*-ee-a.
Philadelphaceae. After Prof. William
M. Carpenter. Evergreen, semi-hardy
shrub.
    californica, kal-i-*for*-ni-ka. Of
    California. Tree Anemone.

Carpinus, kar-*pin*-us. *Carpinaceae.*
The L. name. Deciduous trees.
Hornbeam.
    betulus, bet-ew-lus. *Betula*-like.
    Common Hornbeam.
    caroliniana, ka-ro-lin-ee-*a*-na. Of
    Carolina. American Hornbeam.
    japonica, ja-*pon*-i-ka. Of Japan.
    turczaninowii, tur-cha-ni-*nov*-ee-ee.
    After Nicolai Turczaninow.

Carpinus betulus

Carpobrotus, kar-po-*bro*-tus.
Aizoaceae. From Gk. *karpos* (fruit)
and *brotus* (edible), after the edible
fruit. Semi-hardy, perennial succulent.
    edulis, e-*dew*-lis. Edible (fruit).
    Hottentot Fig.

Carum, *ka*-rum. *Umbelliferae.* From
Gk. *karon.* Biennial and perennial
herbs.
    carvi, kar-vee. The L. name. Caraway.

Carya, *ka*-ree-a. *Juglandaceae.* From
Gk. *karya* (walnut tree). Deciduous
trees. Hickory.
    cordiformis, kor-di-*form*-is. Heart-
    shaped (nut). Bitternut Hickory.
    glabra, gla-bra. Smooth (shoots).
    Pignut Hickory.
    ovata, o-*va*-ta. Ovate. Shagbark
    Hickory.
    tomentosa, to-men-*to*-sa. Woolly
    (shoots). White Heart Hickory.

Caryopteris, ka-ree-*op*-te-ris.
Verbenaceae. From Gk. *karyon* (nut)
and *pteron* (wing), after the winged
fruit. Deciduous sub-shrubs.
    x *clandonensis,* klan-don-*en*-sis.
    From Clandon, Surrey.
    incana, in-*ka*-na. Grey (leaves).

Cassia, *kas*-ee-a. *Leguminosae.* From
Gk. *Kasia.* Tender herbs, trees and
shrubs. Senna.
    alata, a-*la*-ta. Winged (fruit).
    australis, aw-*stra*-lis. Southern.
    floribunda, flo-ri-*bun*-da. Profusely
    flowering.
    marilandica, ma-ri-*land*-i-ka. Of
    Maryland. American Senna.

Cassinia, ka-*sin*-ee-a. *Compositae.*
After Count Henri de Cassini.
Evergreen shrubs.
    fulvida, ful-vi-da. Slightly tawny
    (under the leaf).

Cassiope, ka-*see*-o-pee. *Ericaceae.*
After the mythological Cassiope.
Evergreen shrubs.
    fastigiata, fa-stij-ee-*a*-ta. Erect
    (shoots).
    lycopodioides, lie-ko-po-dee-*oy*-
    deez. *Lycopodium*-like.
    tetragona, tet-ra-*go*-na. Four-angled
    (shoots).
    wardii, ward-ee-ee. After Francis Ward.

Castanea, ka-*sta*-nee-a. *Fagaceae*.
The L. name from Castania, Greece.
Deciduous trees and shrubs.
 sativa, sa-*tee*-va. Cultivated. Sweet
 or Spanish Chestnut.

Catalpa, kat-*al*-pa. *Bignoniaceae*. The
Native American name. Deciduous
trees and shrubs.
 bignonioides, big-no-nee-*oi*-deez.
 Bignonia-like. Indian Bean Tree.
 speciosa, spes-ee-*o*-sa. Showy.

Catananche, ka-ta-*nan*-kee.
Compositae. From Gk. *katananke*.
Perennial herb.
 caerulea, see-*ru*-lee-a. Dark blue.
 Cupid's Dart.

Catharanthus, ka-tha-*ran*-thus.
Apocynaceae. From Gk. *katharos*
(pure) and *anthus* (flower). Tender
annual and perennial herbs.
 roseus, ro-see-us. Rose-coloured.
 Old Maid.

Cattleya, *kat*-lee-a. *Orchidaceae*.
After William Cattley. Evergreen, epiphytic orchids.
 bicolor, bi-ko-lor. Two-coloured.
 bowringiana, bow-ring-gee-*a*-na.
 After John Charles Bowring.
 intermedia, in-ter-*me*-dee-a.
 Between.
 mossiae, mos-ee-ee. After Mrs Moss.
 trianae, tri-*a*-nee. After Dr J. J. Triana.
 warscewiczii, war-sha-*vich*-ee-ee.
 After Warscewicz.

Ceanothus, see-an-*o*-thus.
Rhamnaceae. Gk. name for a spiny
shrub. Deciduous or evergreen trees
and shrubs.
 arboreus, ar-*bo*-ree-us. Tree-like.
 Catalina Ceanothus.
 burkwoodii, burk-*wud*-ee-ee. After

Burkwood.
 dentatus, den-*ta*-tus. Toothed
 (leaves).
 gloriosus, glo-ree-*o*-sus. Glorious.
 papillosus, pa-pil-*lo*-sus. Pimpled.
 prostratus, pros-*tra*-tus. Prostrate.
 repens, ree-penz. Creeping.
 x *veitchianus,* veech-ee-*a*-nus. After
 the Veitch nursery.

Cedrela, *sed*-rel-a. *Meliaceae*.
Diminutive of *Cedrus*. Deciduous tree.
 sinensis, si-*nen*-sis. Of China.

Cedrus, *sed*-rus. *Pinaceae*. The L.
name. Evergreen conifers.
 atlantica, at-*lan*-ti-ka. Of the Atlas
 Mountains.
 glauca, *glaw*-ka. Glaucous (leaves).
 deodara, dee-o-*dar*-a. The Indian
 name. Deodar.
 libani, li-ba-nee. Of Mount Lebanon.
 Cedar of Lebanon.
 sargentii, sar-*jent*-ee-ee. After
 Sargent.

Celastrus, sel-*as*-trus. *Celastraceae*.
From Gk. *kelastros* (evergreen tree).
Deciduous shrubs and climbers.
 orbiculatus, or-bik-ew-*la*-tus. Disc-
 shaped (leaves).

Celmisia, sel-*mis*-ee-a. *Compositae*.
After the mythological Celmisios.
Evergreen perennial herbs.
 coriacea, ko-ree-*a*-see-a. Leathery
 (leaves).
 spectabilis, spek-*ta*-bi-lis.Spectacular.

Celosia, se-*lo*-see-a. *Amaranthaceae*.
From Gk. *keleos* (burning), after the
brilliantly coloured flowers. Semi-
hardy perennial shrubs.
 cristata, kris-*ta*-ta. Crested.
 Cockscomb.
 plumosa, ploo-*mo*-sa. Feathery.

Celtis, *sel*-tis. *Ulmaceae*. Gk. name of another tree. Deciduous trees. Nettle Tree.

australis, aw-*stra*-lis. Southern.
laevigata, lee-vi-*ga*-ta. Smooth (leaves). Sugarberry.
occidentalis, ok-si-den-*ta*-lis. Western. Hackberry.

Centaurea, sent-*aw*-ree-a. *Compositae*. After the mythological Gk. *Kentaur* (Centaur). Annual and perennial herbs.

cyanus, sie-*a*-nus. Dark blue. Cornflower.
dealbata, dee-al-*ba*-ta. Whitened
hypoleuca, hi-po-*loo*-ka. White beneath (leaves).
macrocephala, mak-ro-*sef*-a-la. With a large head.
montana, mon-*ta*-na. Of mountains.
moschata, mos-*ka*-ta. Musk scented. Sweet Sultan.
pulcherrima, pul-*ke*-ri-ma. Very pretty.

Centranthus, sen-*tran*-thus. *Valerianaceae*. From Gk. *kentron*

Centranthus ruber

(spur) and *anthos* (flower), after the spurred flowers. Perennial herb.

ruber, ru-ber. Red. Red Valerian.

Cephalaria, sef-al-*a*-ree-a. *Dipsacaceae*. From Gk. *kephale* (head), after the clustered flowers. Perennial herbs.

alpina, al-*pie*-na. Alpine.
gigantea, ji-*gan*-tee-a. Very large.

Cephalotaxus, sef-a-lo-*tax*-us. *Cephalotaxaceae*. From Gk. *kephale* (head) and *Taxus*. Evergreen conifers. Plum Yew.

fortunei, for-*tewn*-ee-ee. After Robert Fortune. Chinese Plum Yew.
harringtonii, ha-ring-*ton*-ee-ee. After the Earl of Harrington. Japanese Plum Yew.

Cerastium, se-*ras*-tee-um. *Caryophyllaceae*. From Gk. *keras* (horn), after the shape of the seed capsule. Annual and perennial herbs.

alpinum, al-*pie*-num. Alpine.
biebersteinii, bee-ber-*stien*-ee-ee. After Friedrich von Bieberstein.
tomentosum, to-men-*to*-sum. Hairy. Snow in Summer.

Ceratophyllum, ser-at-o-*fil*-um. *Ceratophyllaceae*. From Gk. *keras* (horn) and *phyllon* (leaf). The leaves suggest antlers. Deciduous perennial aquatic herbs. Hornwort.

demersum, dee-*mer*-sum. Growing under water.
submersum, sub-*mer*-sum. Submerged.

Ceratostigma, ser-at-o-*stig*-ma. *Plumbaginaceae*. From Gk. *keras* (horn) and *stigma,* after the horn-like growth on the stigma. Deciduous or evergreen shrubs and herbs.

griffithii, gri-*fith*-ee-ee. After
William Griffith.
plumbaginoides, plum-ba-gi-*noi*-
deez. *Plumbago*-like.
willmottianum, wil-mot-ee-*a*-num.
After Miss Ellen Ann Willmott.

Cercidiphyllum, ser-sid-i-*fil*-lum.
Cercidiphyllaceae. From *Cercis* and
Gk. *phyllon* (leaf), the leaves are
Cercis-like. Deciduous tree.
   japonicum, ja-*pon*-i-kum. Of Japan.

Cercis, *ser*-sis. *Leguminosae.* From
Gk. *kerkis.* Deciduous trees and
shrubs.
   canadensis, kan-a-*den*-sis. Of
   Canada. Redbud.
   siliquastrum, si-li-*kwa*-strum. A
   siliqua-like Judas Tree.

Cereus, *see*-ree-us. *Cactaceae.* From
L. *cereus* (wax taper).
   aethiops, ee-thee-ops. Of unusual
   appearance.
   peruvianus, pe-roo-vee-*a*-nus. Of
   Peru. Peruvian Apple Cactus.

Ceropegia, see-ro-*pee*-jee-a.
Asclepiadaceae. From Gk. *keros* (wax)
and *pege* (fountain), after the waxy
flowers. Tender, semi-evergreen succulents.
   barklyi, *bark*-lee-ee. After Sir Henry
   Barkly.
   woodii, *wud*-ee-ee. After John
   Medley Wood.

Cestrum, *ses*-trum. *Solanaceae.*
Origin unknown. Tender to semi-hardy
deciduous or evergreen shrubs.
aurantiacum, aw-ran-tee-*a*-kum.
Orange.
   elegans, e-le-ganz. Elegant.
   fasciculatum, fas-ik-ew-*la*-tum.
   Clustered.

newellii, new-*el*-ee-ee. After
Newell.
parqui, *par*-kee. The Chilean name.

Chaenomeles, kee-no-*mee*-leez.
Rosaceae. From Gk. *chaina* (gape) and
melon (apple). Deciduous, spiny
shrubs. Flowering Quince.
   cathayensis, ka-thay-*en*-sis. Of
   China.
   japonica, ja-*pon*-i-ka. Of Japan.
   speciosa, spes-ee-*o*-sa. Showy.
   x *superba,* soo-*perb*-a. Superb.

Chamaecereus, kam-ee-*see*-ree-us.
Cactaceae. From Gk. *chamai* (on the
ground) and *Cereus.*
   sylvestrii, sil-*ves*-tree-ee. After Dr
   Philipo Sylvestri. Peanut Cactus.

Chamaecyparis, kam-ee-*sip*-ar-is.
Cupressaceae. From Gk. *chamai* (low
growing) and *kuparissos* (cypress).
Evergreen conifers. False Cypress.
   lawsoniana, law-son-ee-*a*-na. After
   Charles Lawson.
   nootkatensis, noot-ka-*ten*-sis. Of
   Nootka Sound, British Columbia.
   Nootka Cypress.
   obtusa, ob-*tew*-sa. Blunt (leaves).
   Hinoki Cypress.
   pisifera, pi-*si*-fe-ra. Pea-bearing
   (small cones). Sawara Cypress
   thyoides, thoo-*oy*-deez. *Thuja*-like.

Chamaedorea, kam-ee-*do*-ree-a.
Palmae. From Gk. *chamai* (on the
ground) *dorea* (gift). Tender, evergreen
palms.
   elegans, e-le-ganz. Elegant. Parlour
   Palm.

Chamaemelum, kam-ee-*mel*-um.
Compositae. From Gk. *chamai* (on the
ground) and *melon* (apple), after its
apple-like scent and its prostrate habit.

Perennial herb.
*nobile, no-*bi-lee. Notable. Chamomile.

Cheilanthes, ki-*lan-*theez.
Polypodiaceae. From Gk. *cheilos* (lip)
and *anthos* (flower), after its indusium.
Lip Fern.
*distans, dis-*tanz. Widely spaced.
Woolly Rock Fern.
*fragrans, fra-*granz. Fragrant.
lanosa, la-*no-*sa. Woolly. Hairy Lip
Fern.

Cheiranthus, ki-*ranth-*us. *Cruciferae.*
From Gk. *cheir* (hand) and *anthos*
(flower), used in bouquets. Perennial
herbs and sub-shrubs.
allionii, a-lee-*on-*ee-ee. After Allioni.
*cheiri, ki-*ree. Sweet-scented.
Wallflower.
mutabilis, mew-*ta-*bi-lis. Changeable.
speciosa, spes-ee-*o-*sa. Showy.

Chelidonium, kel-i-*do-*nee-um.
Papaveraceae. From Gk. *chelidon*
(swallow). Said to flower as the swallow arrives. Perennial herb.
*majus, ma-*jus. Larger. Greater
Celandine.

Chelidonium majus

Cheiranthus cheiri

Chelone, kel-*o-*nee. *Scrophulariaceae.*
From Gk. *chelone* (turtle). The corolla
suggests a turtle's head. Perennial
herbs. Turtle Head.
*glabra, gla-*bra. Smooth.
lyonii, lie-*on-*ee-ee. After John Lyon.
obliqua, o-*blee-*kwa. Oblique.

Chiastophyllum, kee-as-to-*fil-*lum.
Crassulaceae. From Gk. *chiastos*
(arranged cross-wise) and *phyllon*
(leaf), after the opposing leaves.
Evergreen perennial.
oppositifolium, op-o-sit-i-*fol-*ee-um.
With opposing leaves.

Chimonanthus, kie-mon-*anth-*us.
Calycanthaceae. From Gk. *cheima*
(winter) and *anthos* (flower).
Deciduous shrub.
*praecox, pree-*kox. Early (flowering). Winter Sweet.
*luteus, loo-*tee-us. Yellow.

Chionanthus, kie-on-*anth-*us.
Oleaceae. From Gk. *chion* (snow) and
anthos (flower). Deciduous trees or
shrubs.
retusus, re-*tew-*sus. With a notched
tip (leaves). Chinese Fringe Tree.
virginicus, vir-*jin-*i-kus. Of Virginia.

Chionodoxa, kie-on-o-*dox*-a.
Liliaceae. From Gk. *chion* (snow) and
doxa (glory). They flower in the melt-
ing snow. Bulbous perennial herbs.
Glory of the Snow.
    albescens, al-*bes*-enz. Whitish.
    cretica, *kre*-ti-ka. Of Crete.
    luciliae, loo-*sil*-ee-ee. After Lucile
    Boissier.
    sardensis, sar-*den*-sis. Of Sart,
    Turkey.

Chlidanthus, klid-*anth*-us.
Amaryllidaceae. From Gk. *chlide* (lux-
ury) and *anthos* (flower). Bulbous
perennial herb.
    fragrans, fra-granz. Fragrant.

Chlorophytum, klor-*o*-fi-tum.
Liliaceae. From Gk. *chloros* (green)
and *phyton* (plant). Tender, evergreen
herbs.
    capense, ka-*pen*-see. Of the Cape of
    Good Hope.
    comosum, ko-*mo*-sum. Tufted.
    Spider Plant.

Choisya, *choy*-zee-a. *Rutaceae.* After
Jacques Choisy. Semi-hardy, evergreen
shrub.
    ternata, ter-*na*-ta. In threes (leaflets).
    Mexican Orange Blossom.

Chrysalidocarpus, kris-a-li-do-*karp*-
us. *Palmae.* From Gk. *chrysos* (gold)
and *karpos* (fruit), after the golden
fruit of one species. Tender, evergreen
palm.
    lutescens, loo-*tes*-enz. Yellowish.
    Yellow palm.

Chrysanthemum, kris-*anth*-e-mum.
Compositae. From Gk. *chrysos* (gold)
and *anthos* (flower). Annual and
perennial herbs and evergreen sub-
shrubs.

    alpinum, al-*pie*-num. Alpine.
    carinatum, kar-i-*na*-tum. Keeled.
    Painted Daisy.
    coccineum, kok-*kin*-ee-um. Scarlet.
    Pyrethrum.
    coronarium, ko-ro-*na*-ree-um. Used
    in garlands. Crown Daisy.
    frutescens, froo-*tes*-enz. Shrubby.
    White Marguerite.
    hosmariense, hos-mar-ee-*en*-see. Of
    Beni Hosmar, Morocco.
    indicum, *in*-di-kum. Of India.
    multicaule, mul-ti-*kaw*-lee. Many-
    stemmed.
    segetum, se-je-tum. Of cornfields.
    Corn Marigold
    x *superbum,* soo-*perb*-um. Superb.
    Shasta Daisy.
    weyrichii, way-*rich*-ee-ee. After Dr
    Weyrich.

Chrysanthemum segetum

Chrysogonum, kris-*o*-go-num.
Compositae. From Gk. *chrysos* (gold-
en) and *gonu* (knee), after the yellow
flowers with jointed stems. Perennial
herb.
    virginianum, vir-jin-ee-*a*-num. Of
    Virginia.

Cicerbita, sis-*er*-bit-a. *Compositae.* Italian for a sow-thistle. Perennial herbs.
   alpina, al-*pie*-na. Alpine.
   plumieri, ploo-mee-*e*-ree. After Charles Plumier.

Cichorium, sik-*or*-ee-um. *Compositae.* From the Arabic. Annual, biennial or perennial herbs.
   endivia, en-*di*-vee-a. Endive.
   intybus, *in*-tib-us. Chicory.

Cichorium intybus

Cimicifuga, sim-i-sif-*ew*-ja. *Ranunculaceae.* From L. *cimex* (bug) and *fugo* (to repel), after *C. foetida* (an insect repellent). Perennial herbs. Bugbane.
   americana, a-me-ri-*ka*-na. Of America.
   dahurica, da-*hew*-ri-ka. Of Dahuria, Siberia.
   japonica, ja-*pon*-i-ka. Of Japan.
   racemosa, ra-see-*mo*-sa. With flowers in racemes. Black Snakeroot.

Cionura, see-on-*ewr*-ra. *Asclepiadaceae.* From Gk. *kion* (col-

umn) and *oura* (tail). Deciduous climber.
   erecta, e-*rek*-ta. Erect.

Cissus, *sis*-us. *Vitaceae.* From Gk. *kissos* (ivy), after its climbing habit. Tender evergreen climbers.
   antarctica, an-*tark*-ti-ka. Of Antarctic regions. Kangaroo Vine.
   bainesii, bayn-zee-ee. After Thomas Baines.
   discolor, *dis*-ko-lor. Two-coloured (leaves). Rex-Begonia Vine.
   quadrangularis, kwod-rang-gew-*la*-ris. Four-angled (stems).
   rhombifolia, rom-bi-*fo*-lee-a. With diamond-shaped leaves. Venezuela Treebine.
   striata, stri-*a*-ta. Striped (stems). Miniature Grape Ivy.

Cistus, *sis*-tus. *Cistaceae.* From the Gk. name. Evergreen shrubs. Sun Rose.
   x *aguilari,* a-gwi-*la*-ree. Of Aguilar, Spain.
   albidus, *al*-bi-dus. Whitish. White-leaved Rock Rose.
   x *corbariensis,* kor-ba-ree-*en*-sis. From Corbières, France.
   creticus, *kree*-ti-kus. Of Crete.
   crispus, *kris*-pus. Finely wavy (leaves).
   x *cyprius,* *sip*-ree-us. Of Cyprus.
   ladanifer, la-*da*-ni-fer. Bearing ladanum (myrrh).
   laurifolius, law-ri-*fo*-lee-us. *Laurus*-leaved.
   maculatus, mak-ew-*la*-tus. Spotted (petals).
   populifolius, pop-u-li-*fo*-lee-us. *Populus*-leaved.
   lusitanicus, loo-si-*ta*-ni-kus. From Portugal.
   x *purpureus,* pur-*pur*-ree-us. Purple.
   x *skanbergii,* skan-*berg*-ee-ee. After Skanberg.

Cladanthus, klad-*anth*-us.
Compositae. From Gk. *klados* (branch)
and *anthos* (flower), after the flower
heads on the end of the branches.
Annual herb.
  arabicus, a-*rab*-i-kus. Of Arabia.

Cladrastis, klad-*ras*-tis. *Leguminosae.*
From Gk. *klados* (branch) and *thraus-
tos* (fragile), after its brittle nature.
Deciduous trees.
  lutea, loo-tee-a. Yellow. Yellow
  Wood.
  sinensis, si-*nen*-sis. Of China.

Clarkia, *klar*-kee-a. *Onagraceae.*
After Captain William Clark. Annual
herbs.
  amoena, a-*mee*-na. Pleasant. Satin
  Flower.
  pulchella, pul-*kel*-la. Pretty.

Clematis, *klem*-a-tis. *Ranunculaceae.*
Gk. for climbing plants. Semi-hardy
perennial herbs and deciduous and
evergreen climbers.
  alpina, al-*pie*-na. Alpine.
  armandii, ar-*mond*-ee-ee. After
  Armand David.
  cirrhosa, si-*ro*-sa. With tendrils.
  flammula, *flam*-ew-la. A little flame.
  florida, *flo*-ri-da. Flowering.
  heracleifolia, he-ra-klee-i-*fo*-lee-a.
  Heracleum-leaved.
  integrifolia, in-teg-ri-*fo*-lee-a. With
  entire leaves.
  x *jouiniana,* joo-an-ee-*a*-na. After E.
  Jouin.
  macropetala, mak-ro-*pe*-ta-la. With
  large petals.
  montana, mon-*ta*-na. Of mountains.
  orientalis, o-ree-en-*ta*-lis. Eastern.
  recta, *rek*-ta. Erect.
  rehderiana, red-a-ree-*a*-na. After
  Rehder.
  rubens, *ru*-benz. Red.

  serratifolia, se-ra-ti-*fo*-lee-a. With
  toothed leaves.
  tangutica, tan-*gew*-ti-ka. Of Gansu.
  vitalba, vee-*tal*-ba. White vine. Old
  Man's Beard.

Clematis vitalba

Cleome, klee-*o*-mee. *Capparidaceae.*
Origin unknown. Tender annual herbs
and evergreen shrubs.
  hassleriana, has-la-ree-*a*-na. After
  Emile Hassler. Spider Plant.
  spinosa, spi-*no*-sa. Spiny.

Clerodendrum, kler-o-*den*-drum.
Verbenaceae. From Gk. *kleros*
(chance) and *dendron* (tree), after its
possible healing qualities. Tender to
hardy, evergreen or deciduous trees,
shrubs and climbers.
  bungei, *bun*-jee-ee. After Alexander
  von Bunge. Glory Flower.
  fragrans, fra-granz. Fragrant.
  speciosissimum, spes-ee-o-*sis*-i-
  mum. Most showy. Glory Bower.
  thomsoniae, tom-*son*-ee-ee. After the
  wife of the Rev. W. C. Thomson.
  trichotomum, tri-*ko*-to-mum.
  Branching into three.

Clethra, *kleth*-ra. *Clethraceae.* From
Gk. *klethra* (alder). Deciduous or evergreen trees and shrubs. White Alder.
  alnifolia, al-ni-*fo*-lee-a. Alder-

leaved. Bush Pepper.

arborea, ar-*bo*-ree-a. Tree-like. Lily of the Valley Tree.

barbinervis, bar-bi-*ner*-vis. With bearded veins.

delavayi, del-a-*vay*-ee. After Delavay.

mexicana, mex-i-*ka*-na. Mexican.

Clianthus, kli-*anth*-us. *Leguminosae.* From Gk. *kleos* (glory) and *anthos* (flower), after its colourful flowers. Tender, semi-evergreen climbers.

formosus, for-*mo*-sus. Beautiful. Desert Pea.

puniceus, pew-*ni*-see-us. Reddish-purple. Lobster Claw.

Clivia, *klie*-vee-a. *Amaryllidaceae.* After Lady Charlotte Clive. Tender, evergreen rhizomatous perennial herbs.

nobilis, no-bi-lis. Notable. Greentip Kaffir Lily.

Cobaea, ko-*be*-a. *Polemoniaceae.* After Father Bernardo Cobo. Tender, evergreen or deciduous climber.

scandens, skan-denz. Climbing. Monastery Bells.

Codiaeum, ko-di-*ee*-um. *Euphorbiaceae.* From the Malayan *kodiho.* Tender evergreen shrubs.

variegatum, vair-ee-a-*ga*-tum. Variegated.

pictum, pik-tum. Painted. Croton.

Codonopsis, ko-don-*op*-sis. *Campanulaceae.* From Gk. *kodon* (bell) and *-opsis* (resemblance). Perennial climbers. Bonnet Bellflower.

clematidea, klem-a-*ti*-dee-a. *Clematis*-like.

convolvulacea, kon-vol-vew-*la*-see-a. *Convolvulus*-like.

ovata, o-*va*-ta. Ovate (leaves).

Coelogyne, see-log-*ie*-nee. *Orchidaceae.* From Gk. *koilos* (hollow) and *gyne* (female). Greenhouse orchids.

cristata, kris-ta-ta. Crested (lip).

elata, e-*la*-ta. Tall.

flaccida, fla-si-da. Drooping (racemes).

ochracea, ok-*ra*-see-a. Ochre-yellow

speciosa, spes-ee-*o*-sa. Showy.

Coffea, *kof*-ee-a. *Rubiaceae.* From the Arabic name. Tender, evergreen shrub.

arabica, a-*ra*-bi-ka. Arabian. Arabian Coffee Plant.

Coix, *ko*-ix. *Gramineae.* The Gk. name for a reed-leaved plant. Annual grass.

lacryma-jobi, la-kri-ma-*jo*-bee. Job's tears (seed shape).

Colchicum autumnale

Colchicum, *kol*-chi-kum. *Liliaceae.* Said to have originated from Colchis. Cormous perennial herbs. Autumn Crocus.

agrippinum, ag-ri-*peen*-um. After Agrippina.

autumnale, aw-tum-*na*-lee. Of autumn. Autumn Crocus.
boissieri, bwa-see-*e*-ree. After Pierre Edmund Boissier.
byzantinum, bi-zan-*tee*-num. Of Byzantium (Istanbul).
cilicium, si-*li*-see-um. Of Cilicia, Turkey.
luteum, loo-tee-um. Yellow.
speciosum, spes-ee-*o*-sum. Showy.

Coleus, *ko*-lee-us. *Labiatae.* From Gk. *koleos* (sheath), after the tube which encloses the stamens. Tender perennial herbs and evergreen sub-shrubs.
blumei, bloom-ee-ee. After Carl Ludwig von Blume. Flame Nettle.
thyrsoideus, thur-*soy*-dee-us. Staff-like.

Colletia, ko-*le*-she-a. *Rhamnaceae.* After Philibert Collet. Deciduous, leafless, spiny shrubs.
armata, ar-*ma*-ta. Spiny.
paradoxa, pa-ra-*dox*-a. Unusual.

Collinsia, ko-*linz*-ee-a. *Scrophulariaceae.* After Zaccheus Collins. Annual herbs.
bicolor, bi-ko-lor. Two-coloured.
grandiflora, gran-di-*flo*-ra. Large-flowered. Blue Lips.
violacea, vie-o-*la*-see-a. Violet.

Colquhounia, ko-*hoon*-ee-a. *Labiatae.* After Sir Robert Colquhoun. Semi-evergreen shrubs.
coccinea, kok-*kin*-ee-a. Scarlet.

Columnea, ko-*lum*-nee-a. *Gesneriaceae.* After Fabius Columna. Tender, evergreen climbers. Costa Rica.
gloriosa, glo-ree-*o*-sa. Glorious.
linearis, lin-ee-*a*-ris. Narrow (leaves).
microphylla, mie-kro-*fil*-a. With small leaves.

Convallaria majalis

Colutea, ko-*loo*-tee-a. *Leguminosae.* From Gk. *kolutea.* Deciduous shrubs and trees.
arborescens, ar-bo-*res*-enz. Tree-like. Bladder Senna.
orientalis, o-ree-en-*ta*-lis. Eastern.

Commelina, kom-el-*en*-a. *Commelinaceae.* After Johan and Caspar Commelin. Semi-hardy perennial herbs. Day Flower.
coelestis, see-*les*-tis. Sky-blue.
erecta, e-*rek*-ta. Erect.
tuberosa, tew-be-*ro*-sa. Tuberous.

Conophytum, kon-o-*fie*-tum. *Aizoaceae.* From Gk. *konos* (cone) and *phyton* (plant), after its inverted cone shape.
albescens, al-*bes*-enz. Whitish.
bilobum, bi-*lo*-bum. Two-lobed.
meyeri, may-a-ree. After F. N. Meyer

Consolida, kon-*sol*-i-da. *Ranunculaceae.* From L. *consolida* (make whole), after its healing properties. Annual herbs. Larkspur.
ambigua, am-*big*-ew-a. Doubtful.
regalis, ree-*ga*-lis. Regal.

Convallaria, kon-va-*la*-ree-a.
Liliaceae. From L. *convallis* (valley).
Rhizomatous perennial herbs.
   majalis, ma-*ja*-lis. Flowering in
May. Lily of the Valley.

Convolvulus, kon-*vol*-vew-lus.
Convolvulaceae. From L. *convolva*
(entwine). Annual and perennial herbs,
shrubs and climbers.
   althaeoides, al-thee-*oi*-deez.
Althaea-like.
   cneorum, nee-*o*-rum. Olive-like
shrub. Silver Bush.
   purpureus, pur-*pur*-ree-us. Purple.
   tricolor, *tri*-ko-lor. Three-coloured.

Coprosma, kop-*ros*-ma. *Rubiaceae.*
From Gk. *kopros* (dung) and *osme*
(smell), after the smell when bruised.
Tender to semi-hardy evergreen trees
and shrubs.
   lucida, *loo*-si-da. Glossy (leaves).
   petriei, *pet*-ree-ee. After Donald
Petrie.
   repens, *ree*-penz. Creeping.

Cordyline, kor-*di*-lie-nee. *Agavaceae.*
From Gk. *kordyle* (club), after the
fleshy roots. Tender to semi-hardy,
evergreen trees and shrubs.
   australis, aw-*stra*-lis. Southern.
Cabbage Palm.
   indivisa, in-di-*vee*-sa. Undivided, it
is usually single-stemmed.
   rubra, *rub*-ra. Red.
   stricta, *strik*-ta. Upright.
   terminalis, ter-min-*a*-lis. Terminal.

Coreopsis, ko-ree-*op*-sis. *Compositae.*
From Gk. *koris* (bug) and -*opsis*
(resemblance), after the tick-like seeds.
Annual and perennial herbs. Tickseed.
   auriculata, aw-rik-ew-*la*-ta. With
auricles (leaves).
   grandiflora, gran-di-*flo*-ra. Large-

flowered.
   lanceolata, lan-see-o-*la*-ta. Spear-
shaped (leaves).
   tinctoria, tink-*to*-ree-a. Used in dyeing.
   verticillata, ver-ti-sil-*a*-ta. Whorled
(leaves).

Coriaria, ko-ree-*a*-ree-a.
Coriariaceae. From L. *corium*
(leather), after its use in tanning. Semi-
hardy deciduous shrubs and sub-
shrubs.
   japonica, ja-*pon*-i-ka. Of Japan.
   nepalensis, ne-pa-*len*-sis. Of Nepal.
   terminalis, ter-mi-*na*-lis. Terminal.
   xanthocarpa, zanth-o-*kar*-pa. With
yellow fruits.

Cornus, *kor*-nus. *Cornaceae.* L. name
for *C. mas.* Deciduous shrubs and
evergreen trees. Dogwood.
   alba, *al-ba.* White (fruit). Red-
barked Dogwood.
   alternifolia, al-ter-ni-*fo*-lee-a. With
alternate leaves. Green Osier.
   canadensis, kan-a-*den*-sis. Of
Canada.

Cornus sanguinea

capitata, kap-i-*ta*-ta. A dense head.
florida, flo-ri-da. Free-flowering.
kousa, koo-sa. The Japanese name.
chinensis, chin-*en*-sis. Of China.
macrophylla, mak-ro-*fil-a*. With large leaves.
mas, mas. Male. Cornelian Cherry.
nuttallii, nu-*tal*-ee-ee. After Thomas Nuttall. Pacific Dogwood.
sanguinea, sang-*gwin*-ee-a. Red (autumn colour). Common Dogwood, Europe.
stolonifera, sto-lo-*ni*-fe-ra. Bearing stolons. American Dogwood.

Corokia, ko-*ro*-kee-a. *Cornaceae.* From the Maori *korokia.* Evergreen shrubs.
cotoneaster, ko-ton-ee-*a*-ster. *Cotoneaster*-like.
x *virgata,* vir-*ga*-ta. Twiggy.

Coronilla, ko-ro-*nil*-la. *Leguminosae.* Diminutive of L. *corona* (crown), after the umbrels. Hardy to semi-hardy deciduous or evergreen shrubs.
glauca, glaw-ka. Glaucous (leaves).
valentina, val-en-*teen*-a. Of Valencia, Spain.

Correa, ko-*ree*-a. *Rutaceae.* After José Francesco Correa de Serra. Tender to semi-hardy evergreen shrubs.
alba, al-ba. White (flowers).
reflexa, re-*flex*-a. Bent backwoods (corolla lobes).

Cortaderia, kor-ta-*der*-ee-a. *Gramineae.* From the Argentinian name. Perennial grasses.
selloana, sel-o-*a*-na. After Sellow. Pampas Grass.

Corydalis, ko-*ri*-da-lis. *Papaveraceae.* The Gk. name for a lark, which the spur resembles. Annual and perennial herbs.

Corydalis bulbosa

bulbosa, bul-*bo*-sa. Bulbous (tuber).
cashmeriana, kash-me-ree-*a*-na. Of Kashmir.
cheilanthifolia, kie-lanth-i-*fo*-lee-a. With *Cheilanthus*-like leaves.
lutea, loo-tee-a. Yellow.
nobilis, no-bi-lis. Noble.
solida, so-li-da. Solid (tuber).

Corylopsis, ko-ri-*lop*-sis. *Hamamelidaceae.* From Gk. *Corylus* and *-opsis* (resemblance), it is similar to *Corylus.* Deciduous trees and shrubs.
glabrescens, gla-*bres*-enz. Nearly smooth (leaves).
pauciflora, paw-si-*flo*-ra. With few flowers (spikes).
sinensis, si-*nen*-sis. Of China.
spicata, spi-*ka*-ta. With flowers in spikes.
willmottiae, wil-*mot*-ee-ee. After Ellen Willmott.

Corylus, ko-*ril*-us. *Corylaceae.* From Gk. *korylos.* Deciduous shrubs and trees. Hazel.
avellana, a-ve-*la*-na. Of Avella. Hazelnut.
contorta, kon-*tor*-ta. Twisted (shoots). Corkscrew Hazel.
colurna, ko-*lur*-na. The classical name. Turkish Hazel
maxima, max-i-ma. Larger. Filbert.

Cosmos, *kos*-mos. *Compositae*. From Gk. *kosmos* (beautiful). Annual and perennial tuberous herbs.
 bipinnatus, bi-pin-*a*-tus. With bi-pinnate leaves.

Cotinus, *ko*-ti-nus. *Anacardiaceae*. From Gk. *kotinos* (olive). Deciduous trees and shrubs.
 coggygria, ko-*jig*-ree-a. From *kokkugia* the Gk. name. Smoke Tree.
 obovatus, ob-o-*va*-tus. Inverted ovate (leaves).

Cotoneaster, ko-ton-ee-*as*-ter. *Rosaceae*. From L. *cotoneum* (quince) and *-aster* (likeness). Deciduous and evergreen trees and shrubs.
 affinis, a-*fee*-nis. Related to.
 amoenus, a-*mee*-nus. Pleasant.
 bacillaris, ba-si-*la*-ris. Staff-like.
 congestus, con-*jes*-tus. Congested.
 divaricatus, di-va-ri-*ka*-tus. With spreading branches.
 floribundus, flo-ri-*bun*-dus. Profusely flowering.
 franchetii, fran-*shet*-ee-ee. After Adrien Franchet.

Corylus avellana

frigidus, fri-ji-dus. Growing in cold regions.
horizontalis, ho-ri-zon-*ta*-lis. Horizontal.
lacteus, lak-tee-us. Milky (flowers).
microphyllus, mie-kro-*fil*-lus. Small-leaved.
multiflorus, mul-ti-*flo*-rus. Many-flowered.
praecox, pree-kox. Very early (fruit ripening).
prostratus, pros-*tra*-tus. Prostrate.
salicifolius, sa-lis-i-*fo*-lee-us. *Salix*-leaved.
simonsii, sie-*monz*-ee-ee. After Mr Simons.
sternianus, stern-ee-*a*-nus. After Sir Frederick Stern.

Cotula, *ko*-tew-la. *Compositae*. From Gk. *kotula* (small cup). The leaves form a small cup. Annual and perennial herbs. Brass buttons.
 atrata, a-*tra*-ta. Black (flowers).
 barbata, bar-*ba*-ta. Bearded (stems).
 coronopifolia, ko-ro-no-pi-*fo*-lee-a. *Coronopus*-leaved.

Cotyledon, ko-ti-*le*-don. *Crassulaceae*. From Gk. *kotyle* (small cup), after the cupped leaves. Tender, evergreen succulent shrubs.
 ladysmithensis, lay-dee-smith-*en*-sis. From Ladysmith, S. Africa.
 orbiculata, or-bik-ew-*la*-ta. Disc-shaped (leaves).
 paniculata, pan-ik-ew-*la*-ta. With flowers in panicles.
 reticulata, ree-tik-ew-*la*-ta. Net-veined.
 undulata, un-dew-*la*-ta. Wavy-edged (leaves). Silver Crown.

Crambe, *kram*-bee. *Cruciferae*. Gk. name for cabbage. Annual and perennial herbs.

cordifolia, kor-di-*fo*-lee-a. With heart-shaped leaves.
maritima, ma-*ri*-ti-ma. Growing near the sea. Sea Kale.

Crassula, *kras*-ew-la. *Crassulaceae.* From L. *crassus* (thick), after the fleshy leaves. Tender, perennial succulents and evergreen shrubs.
arborescens, ar-bo-*res*-enz. Tree-like. Chinese Jade.
argentea, ar-*jen*-tee-a. Silvery.
brevifolia, brev-i-*fo*-lee-a. Short-leaved.
cooperi, *koo*-pa-ree. After Thomas Cooper.
falcata, fal-*sa*-ta. Sickle-shaped (leaves).
lactea, *lak*-tee-a. Milky (flowers). Tailor's Patch.
lycopodioides, lie-ko-po-dee-*oi*-deez. *Lycopodium*-like.
sarcocaulis, sar-ko-*kaw*-lis. Fleshy-stemmed.
schmidtii, *shmit*-ee-ee. After E. Schmidt.
socialis, so-see-*a*-lis. Growing in colonies.

Crataegus, kra-*te*-gus. *Rosaceae.* From Gk. *kratos* (strength), after its hard wood. Deciduous trees. Hawthorn.
crus-galli, kroos-*gal*-ee. A cock's spur. Cockspur Thorn.
flava, *fla*-va. Yellow.
laciniata, la-sin-ee-*a*-ta. Deeply cut (leaves).
laevigata, lee-vi-*ga*-ta. Smooth (leaves).
monogyna, mon-*o*-gi-na. With one pistil. Hawthorn.
orientalis, o-ree-en-*ta*-lis. Eastern.
prunifolia, proon-i-*fo*-lee-a. *Prunus*-leaved.
tanacetifolia, ta-na-set-i-*fo*-lee-a. *Tanacetum*-leaved.

Crepis, *kre*-pis. *Compositae.* From Gk. *krepis* (sandal). Annual, biennial and perennial herbs. Dandelion.
aurea, *aw*-ree-a. Golden.
incana, in-*ka*-na. Grey-hairy (leaves). Pink dandelion.
rubra, *rub*-ra. Red.

Crinodendron, krin-o-*den*-dron. *Elaeocarpaceae.* From Gk. *krinon* (lily) and *dendron* (tree). Semi-hardy, evergreen trees and shrubs.
hookerianum, huk-a-ree-*a*-num. After Hooker. Chile Lantern Tree.

Crinum, *krie*-num. *Amaryllidaceae.* From Gk. *krinon* (lily). Semi-hardy, bulbous perennial herbs.
americanum, a-me-ri-*ka*-num. Of America.
asiaticum, a-see-*a*-ti-kum. Of Asia. Poison Blue.
bulbispermum, bul-bee-*sperm*-um. With bulbous seeds.
campanulatum, kam-pan-ew-*la*-tum. Bell-shaped
moorei, *mor*-ree-ee. After Moore.
x *powellii,* *pow*-el-ee-ee. After C. Baden-Powell.

Crataegus monogyna

Crocosmia, kro-*kos*-mee-a. *Iridaceae.*
From Gk. *krokos* (saffron) and *osme*
(smell). The dried flowers smell of
saffron. Cormous perennial herbs.
Montbretia.
  aurea, aw-ree-a. Golden.
  masoniorum, may-son-ee-*or*-rum.
  After Canon G. E. and Miss Mason.
  paniculata, pa-nik-ew-*la*-ta. With
  flowers in panicles.

Crocus, *kro*-kus. *Iridaceae.* From Gk.
krokos (saffron). Cormous, perennial
herbs.
  angustifolius, an-gus-ti-*fo*-lee-us.
  Narrow-leaved.
  banaticus, ba-*na*-ti-kus. Of Banat,
  Romania.
  biflorus, bi-*flo*-rus. Two-flowered.
  Scotch Crocus.
  cancellatus, kan-sel-*a*-tus. Latticed
  (corm tunic).
  chrysanthus, kris-*anth*-us. Golden-
  flowered.
  dalmaticus, dal-*ma*-ti-kus. Of
  Dalmatia.
  etruscus, e-*troos*-kus. Of Tuscany.
  flavus, fla-vus. Yellow.
  imperati, im-pe-*ra*-tee. After
  Ferrante Imperato.
  korolkowii, ko-rol-*kow*-ee-ee. After
  General Korolkow.
  kotschyanus, kot-shee-*a*-nus. After
  Theodor Kotschy.
  laevigatus, lee-vi-*ga*-tus. Smooth
  (corm tunic).
  longiflorus, long-i-*flo*-rus. Long-
  flowered.
  medius, me-dee-us. Intermediate.
  minimus, min-i-mus. Smaller.
  niveus, niv-ee-us. Snow white.
  nudiflorus, new-di-*flo*-rus. Flowers
  arriving before the leaves.
  olivieri, o-liv-ee-*e*-ree. After
  Guillaume Antoine Oliver.
  pulchellus, pul-*kel*-lus. Pretty.

Crocus speciosus

  sativus, sa-*tee*-vus. Cultivated.
  Saffron.
  serotinus, se-*ro*-ti-nus. Late flowering.
  salzmannii, saltz-*man*-ee-ee. After
  Philip Salzmann.
  sieberi, see-ba-ree. After France
  William Sieber.
  speciosus, spes-ee-*o*-sus. Showy.
  tommasinianus, tom-a-see-nee-*a*-
  nus. After Muzio de Tommasini.
  vernus, ver-nus. Of spring (flowering). Dutch Crocus.

Crossandra, kros-*an*-dra.
Acanthaceae. From Gk. *krossos*
(fringe) and *aner* (male), after the
fringed anthers. Tender, evergreen
perennial herbs and shrubs.
  infundibuliformis, in-fun-dib-ew-lee-
  form-is. Trumpet-shaped (flowers).
  Firecracker Flower.
  nilotica, ni-*lo*-ti-ka. Of the Nile
  Valley.

Cryptanthus, krip-*tanth*-us.
Bromeliaceae. From Gk. *krypto* (hide)
and *anthos* (flower), after the hidden

flowers. Tender, evergreen perennial
herbs. Earth Star.
 acaulis, a-*kaw*-lis. Stemless. Starfish
 Plant.
 bivittatus, bi-vi-*ta*-tus. With two
 stripes (leaves).
 bromelioides, brom-ee-lee-*oi*-deez.
 Bromelia-like. Rainbow Star.
 zonatus, zo-*na*-tus. Banded (leaves).
 Zebra Plant.

Cryptogramma, krip-to-*gram*-a.
Cryptogrammaceae. From Gk. *krypto*
(hide) and *gramma* (line), after the
hidden spore cases. Deciduous or evergreen ferns. Parsley Fern.
 crispa, kris-pa. Curly (fronds).

Cryptomeria, krip-to-*mer*-ee-a.
Toxodiaceae. From Gk. *krypo* (hide)
and *meris* (part), after the hidden flowers. Japanese Cedar.
 japonica, ja-*pon*-i-ka. Of Japan.
 elegans, e-le-ganz. Elegant.
 vilmoriniana, vil-mo-rin-ee-*a*-na.
 After M. de Vilmorin.

Ctenanthe, sten-*an*-thee.
Marantaceae. From Gk. *kteinos*
(comb) and *anthos* (flower), after the
arrangement of the bracts. Tender,
evergreen perennial herbs.
 lubbersiana, lub-erz-ee-*a*-na. After
 C. Lubbers.
 oppenheimiana, o-pan-hiem-ee-*a*-na.
 After Edouard Oppenheim.

Cunninghamia, kun-ing-*ham*-ee-a.
Taxodiaceae. After James
Cunningham. Evergreen conifer.
 lanceolata, lan-see-o-*la*-ta. Spear-
 shaped (leaves).

Cuphea, *kew*-fee-a. *Lythraceae.* From
Gk. *kyphos* (curved), after the shaped
seed capsule. Tender, annual and

perennial herbs and shrubs.
 cyanea, sie-*an*-ee-a. Blue.
 hyssopifolia, hi-sop-i-*fo*-lee-a.
 Hyssopus-leaved.
 ignea, ig-nee-a. Glowing (red calyx).
 Cigar Plant.

Cupressus, kew-*pres*-us.
Cupressaceae. L. name for *C. sempervirens.* Evergreen conifers. Cypress.
 arizonica, a-ri-*zo*-ni-ka. Of Arizona.
 cashmeriana, kash-me-ree-*a*-na. Of
 Kashmir.
 glabra, gla-bra. Smooth (bark).
 lusitanica, loo-si-*ta*-ni-ka. Of
 Portugal.
 macrocarpa, mak-ro-*kar*-pa. Large-
 fruited. Monterey Cypress.
 sempervirens, sem-per-*vi*-renz.
 Evergreen. Italian Cypress.

Cyananthus, sie-a-*nanth*-us.
Campanulaceae. From Gk. *kyanos*
(blue) and *anthos* (flower). Perennial
herbs. Trailing Bellflower.
 lobatus, lo-*ba*-tus. Lobed (leaves).
 microphyllus, mie-kro-*fil*-lus. Small-
 leaved.

Cyanotis, sie-a-*no*-tis.
Commelinaceae. From Gk. *kyanos*
(blue) and *ous* (ear), after the petal
shape and colour. Tender, perennial
herbs.
 kewensis, kew-*en*-sis. Of Kew.
 Teddy Bear Plant.
 somaliensis, so-ma-lee-*en*-sis. Of
 Somalia. Pussy Ears.

Cyclamen, *sik*-la-men. *Primulaceae.*
The Gk. name. Tender to hardy, perennial herbs. Persian Violet.
 cilicium, si-*li*-see-um. Of Cilicia,
 Turkey.
 coum, ko-um. Of Kos.
 creticum, kre-ti-kum. Of Crete.

Cyclamen purpurascens

cyprium, sip-ree-um. Of Cyprus.
graecum, gree-kum. Of Greece.
hederifolium, he-de-ri-*fo*-lee-um.
Hedera-leaved.
persicum, per-si-kum. Of Persia
(Iran).
purpurascens, pur-pur-*ras*-enz.
Purplish (flowers).
repandum, re-*pan*-dum. Wavy-margined (leaves).
rohlfsianum, rolfs-ee-*a*-num. After
Rohlfs.

Cymbalaria, sim-ba-*la*-ree-a.
Scrophulariaceae. From Gk. *kymbalon*
(cymbal), after the leaf shape.
Perennial herbs.
muralis, mew-*ra*-lis. Growing on
walls. Ivy-leaved Toadflax.

Cymbidium, sim-*bid*-ee-um.
Orchidaceae. From Gk. *kymbe* (boat),
after the hollowed lip. Greenhouse
orchids.
eburneum, e-*burn*-ee-um. Ivory.
dayanum, day-*a*-num. After John
Day.
elegans, e-le-ganz. Elegant.
giganteum, ji-*gan*-tee-um. Very
large.

grandiflorum, gran-di-*flo*-rum. With
large flowers.
pendulum, pen-dew-lum. Pendulous.
pumilum, pew-mi-lum. Dwarf.
tigrinum, ti-*gree*-num. Striped like a
tiger.
tracyanum, tray-sie-*a*-num. After
Henry Tracy.

Cynara, sin-*a*-ra. *Compositae.* The L.
name. Perennial herbs.
cardunculus, kar-dun-*kew*-lus.
Thistle-like. Cardoon.
scolymus, sko-li-mus. Scolymus-like.
Globe Artichoke.

Cynoglossum, si-no-*glos*-um.
Boraginaceae. From Gk. *kyon* (dog)
and *glossum* (tongue), after the leaf
texture. Annual, biennial or perennial
herbs.
amabile, a-*ma*-bi-lee. Beautiful.
nervosum, ner-*vo*-sum. Distinctly
veined (leaves).

Cyperus, sie-*pe*-rus. *Cyperaceae.* Gk.
for a sedge plant. Tender, perennial
herbs. Sedge.
albostriatus, al-bo-stri-*a*-tus. White-
striped leaf veins.
longus, long-us. Long (stems).
Galingale.
papyrus, pa-*pi*-rus. Paper. Papyrus.

Cypripedium, sip-ree-*pee*-dee-
um.*Orchidaceae.* From Gk. *kypris*
(Venus) and *pedilon* (slipper), after the
shape of the flower. Hardy orchids.
Slipper Orchid.
calceolus, kal-*see*-o-lus. A small
shoe.
reginae, ree-*jeen*-ee. Of the Queen.

Cyrtanthus, ser-*tanth*-us.
Amaryllidaceae. From Gk. *kyrtos*
(arched) and *anthos* (flower). Tender

to hardy, bulbous perennial herbs. Fire
Lily.

angustifolius, an-gust-i-*fo*-lee-us.
Narrow-leaved. Fire Lily.

mackenii, ma-*ken*-ee-ee. After Mark
John Macken. Ifafa Lily.

parviflorus, par-vi-*flo*-rus. Small-
flowered.

sanguineus, sang-*win*-ee-us. Blood-
red (flowers).

Cystopteris, sis-*top*-te-ris.
Polypodiaceae. From Gk. *kystos* (blad-
der) and *pteris* (fern). Deciduous ferns.
Bladder Fern.

bulbifera, bul-*bi*-fe-ra. Bulb-bearing.
Berry Bladder Fern.

dickeana, dik-ee-*a*-na. After Prof.
George Dickie.

fragilis, fra-ji-lis. Brittle. Brittle Fern.

Cytisus, *si*-ti-sus. *Leguminosae.* From
Gk. *kytisos.* Deciduous or evergreen
shrubs. Broom.

albus, al-bus. White.

battandieri, ba-ton-dee-*e*-ree. After
Jules Aime Battandier.

x *beanii, been*-ee-ee. After William
Jackson Bean.

x *kewensis,* kew-*en*-sis. Of Kew.

nigricans, nig-ri-kanz. Blackish.

purpureus, pur-*pur*-ree-us. Purple.

racemosus, ra-see-*mo*-sus. With
flowers in racemes.

D

Daboecia, da-bo-*ee*-see-a.
Ericaceae. After St. Dabeoc.
Evergreen shrubs.
  azorica, a-*zo*-ri-ka. Of the Azores.
  bicolor, bi-ko-lor. Two-coloured.
  cantabrica, kan-*ta*-bri-ka. Of
  Cantabria, Spain.
  praegerae, pray-ga-ree. After Mrs
  Praeger.
  x *scotica, sko*-ti-ka. Of Scotland.

Dactylorrhiza, dak-til-o-*ree*-za.
Orchidaceae. From Gk. *dactylos* (finger) and *rhiza* (root), with finger-like
tubers. Hardy orchids.
  elata, e-*la*-ta.Tall.
  foliosa, fo-lee-*o*-sa. Leafy.
  fuchsii, few-she-ee. After Fuchs.
  Common Spotted Orchid.
  incarnata, in-kar-*na*-ta. Pink.
  Meadow Orchid.
  majalis, ma-*ja*-lis. May-flowering.

Dahlia, *day*-lee-a. *Compositae.* After
Dr Anders Dahlt. Tender and semi-
hardy tuberous perennials.
  coccinea, kok-*kin*-ee-a. Scarlet.
  hortensis, hor-*ten*-sis. Of gardens.
  imperialis, im-peer-e-*a*-lis.
  Powerful.
  pinnata, pin-*a*-ta. Pinnate.

Daphne, *daf*-nee. *Thymelaeaceae.* The
Gk. name for *Laurus nobilis.*
Deciduous and evergreen or semi-
evergreen shrubs.
  alpina, al-*pie*-na. Alpine.
  arbuscula, ar-*bus*-kew-la. Like a
  dwarf tree.
  aureo-marginata, aw-ree-o-mar-ji-
  na-ta. Gold-margined (leaves).

Daphne laureola

  bholua, bo-*loo*-a. From the native
  name, Bholu Swa.
  blagayana, bla-gay-*a*-na. After
  Count Blagay.
  x *burkwoodii,* burk-*wud*-ee-ee. After
  Albert Burkwood.
  collina, ko-*lee*-na. Growing on hills.
  giraldii, ji-*ral*-dee-ee. After
  Giuseppe Giraldi.
  laureola, law-*ree*-o-la. A little laurel.
  Spurge Laurel.
  mezereum, me-*ze*-ree-um. From *mezereon.* Mezereon.
  odora, o-*do*-ra. Fragrant.
  petraea, pe-*tree*-a. Growing on
  rocks.
  retusa, re-*tew*-sa. With a notched
  apex (leaves).
  tangutica, tan-*gew*-ti-ka. Of Gansu.

Daphniphyllum, daf-nee-*fil*-lum.
Daphniphyllaceae. From Gk. *daphne*

(laurel) and *phyllon* (leaf). Evergreen trees and shrubs.
  macropodum, ma-*kro*-po-dum. With a large stalk.

Datura, da-*tewr*-ra. *Solanaceae.* From a native name. Tender, semi-evergreen and evergreen trees and shrubs.
  arborea, ar-*bo*-ree-a. Tree-like. Angels' Trumpets.
  aurea, aw-ree-a. Golden.
  candida, kan-di-da. White.
  sanguinea, san-*gwin*-ee-a. Blood-red (corolla).
  suaveolens, swa-*vee*-o-lenz. Sweetly scented.

Daucus, *dow*-kus. *Umbelliferae.* The L. name. Biennial herb.
  carota, ka-*rot*-a. Red-rooted. Wild Carrot.
  sativus, sa-*tee*-vus. Carrot.

Davallia, da-*val*-ee-a. *Davalliaceae.* After Edmond Davall. Tender, ever-green or semi-evergreen ferns.
  canariensis, ka-na-ree-*en*-sis. Of the Canary Islands. Deer's Foot Fern.
  mariesii, ma-*reez*-ee-ee. After Charles Maries. Squirrel's Foot Fern.

Davidia, da-*vid*-ee-a. *Davidiaceae.* After Armand David. Deciduous tree.
  involucrata, in-vo-loo-*kra*-ta. With an involucre (bracts). Dove Tree, Handkerchief Tree.

Decaisnea, de-*kayz*-nee-a. *Lardizabalaceae.* After Joseph Decaisne. Deciduous shrub.
  fargesii, far-*geez*-ee-ee. After Père Paul Guillaume Farges.

Decumaria, dek-ew-*ma*-ree-a. *Hydrangeaceae.* From L. *decimus*

(ten), the flower parts are in tens. Evergreen or deciduous climbers.
  barbara, bar-ba-ra. Foreign.
  sinensis, sin-*en*-sis. Of China.

Delphinium, del-*fin*-ee-um. *Ranunculaceae.* From the Gk. *delphis* (dolphin). Annual and perennial herbs.
  ajacis, aj-*a*-kis. After Ajax.
  cardinale, kar-di-*na*-lee. Scarlet.
  chinense, chin-*en*-see. Of China.
  elatum, e-*la*-tum. Tall.
  exaltatum, ex-al-*ta*-tum. Very tall.
  grandiflorum, gran-di-*flo*-rum. Large-flowered.
  nudicaule, new-di-*kaw*-lee. With a bare stem.
  tatsienense, tat-see-en-*en*-see. Of Tatsienlu.
  zalil, za-lil. The Afghan name.

Delphinium ajacis

Dendrobium, den-*dro*-bee-um. *Orchidaceae.* From Gk. *dendron* (tree) and *bios* (life), growing in trees. Greenhouse orchids.
  aphyllum, a-*fil*-lum. Leafless.
  bigibbum, bi-*gib*-um. Two-humped.

densiflorum, dens-i-*flo*-rum.
Densely-flowered.
fimbriatum, fim-bree-*a*-tum. Fringed.
infundibulum, in-fun-*dib*-ew-lum.
Funnel-shaped (flowers).
moschatum, mos-*ka*-tum. Musk-
scented.
nobile, no-bi-lee. Noble.
primulinum, prim-ew-*leen*-um.
Primrose-coloured.
williamsonii, wil-yam-*son*-ee-ee.
After Mr W. J. Williamson.

Dendrochilum, den-dro-*keel*-um.
Orchidaceae. From Gk. *dendron* (tree)
and *cheilos* (lip). Evergreen greenhouse orchids.
cobbianum, cob-ee-*a*-num. After
Walter Cobb.
filiforme, fee-lee-*form*-ee. Thread
like.
glumaceum, gloo-*ma*-see-um. With
chaffy bracts. Silver Chain.

Dennstaedtia, den-*stet*-ee-a.
Dennstaedtiaceae. After August
Wilhelm Dennstedt. Deciduous or
semi-evergreen ferns.
punctilobula, punk-tee-*lob*-ew-la.
With dotted lobules. Hay-scented
Fern.

Dendromecon, den-dro-*mee*-kon.
Papaveraceae. From Gk. *dendron*
(tree) and *mecon* (poppy). Semi-hardy
evergreen shrubs.
rigida, ri-ji-da. Rigid (leaves).

Desfontainea, des-fon-*tay*-nee-a.
Potaliaceae. After Rene Louiche
Desfontaines. Semi-hardy, evergreen
shrubs.
spinosa, spi-*no*-sa. Spiny (leaves).

Deutzia, *doytz*-ee-a. *Philadelphaceae.*
After Johann van der Deutz.

Deciduous shrubs.
compacta, com-*pak*-ta. Compact
(inflorescence).
corymbiflora, ko-rim-bee-*flo*-ra.
With flowers in corymbs.
x *elegantissima,* e-le-gan-*tis*-i-ma..
Most elegant.
gracilis, gra-si-lis. Graceful. .
longifolia, long-i-*fo*-lee-a. With long
leaves.
x *magnifica,* mag-*ni*-fi-ka.
Magnificent.
pulchra, pul-kra. Pretty.
scabra, ska-bra. Rough (leaves).
setchuanensis, sech-wan-*en*-sis. Of
Sichuan, China.
veitchii, veech-ee-ee. After the
Veitch nursery.

Dianella, dee-a-*nel*-la. *Liliaceae.*
Diminutive of Diana. Evergreen perennial herbs. Flax Lily.
caerulea, see-*ru*-lee-a. Dark blue.
intermedia, in-ter-*me*-dee-a.
Intermediate.
nigra, nig-ra. Black.
tasmanica, taz-*man*-i-ka. Of
Tasmania.

Dianthus, dee-*anth*-us.
Caryophyllaceae. From Gk. *Di* (of
Zeus or Jove) and *anthos* (flower).
Annual, biennial and perennial herbs.
Carnation, Pink.
x *allwoodii,* awl-*wud*-ee-ee. After M.
C. W. Allwood.
alpinus, al-*pie*-nus. Alpine.
barbatus, bar-*ba*-tus. Bearded
(petals). Sweet William.
callizonus, ka-lee-*zon*-us.
Beautifully zoned (petals).
carthusianorum, kar-thew-zee-a-*nor*-
rum. Of the Carthusian Monks.
caryophyllus, ka-ree-o-*fil*-lus.
Smelling of cloves. Carnation, Clove
Pink.

Dianthus deltoides

chinensis, chin-*en*-sis. Of China.
Indian Pink.
deltoides, del-*toi*-deez. Triangular
(petals). Maiden Pink.
glacialis, gla-see-*a*-lis. Growing
near glaciers. Glacier Pink.
haematocalyx, hee-ma-to-*ka*-lix.
With a blood-red calyx.
myrtinervius, mur-tee-*ner*-vee-us.
With *Myrtus*-like veins.
pavonius, pa-*vo*-nee-us. Peacock
blue.
petraeus, pe-*tree*-us. Growing on
rocks.
noeanus, no-ee-*a*-nus. After
Friedrich Wilhelm Noe.
plumarius, ploo-*ma*-ree-us. Plumed
(fringed petals).
superbus, soo-*perb*-us. Superb.
Fringed Pink.

Diascia, dee-*as*-ee-a.
Scrophulariaceae. From Gk. *di* (two)
and *askos* (sac). Annual and perennial
herbs.
barberiae, bar-*be*-ree-ee. After Mrs
Barber. Twinspur.
cordata, kor-*da*-ta. Heart-shaped
(leaves).
rigescens, ri-*ges*-enz. Somewhat
rigid.

Dicentra, di-*sen*-tra. *Fumariaceae.*
From Gk. *di* (two) and *kentron* (spur).
Perennial herbs.
cucullaria, kuk-ew-*la*-ree-a. Hooded
flowers. Dutchman's Breeches.
eximia, ex-*im*-ee-a. Distinguished.
formosa, for-*mo*-sa. Beautiful.
spectabilis, spek-*ta*-bi-lis.
Spectacular. Bleeding Heart.

Dicksonia, dik-*son*-ee-a.
Dicksoniaceae. After James Dickson.
Tender, evergreen and semi-evergreen
tree ferns.
antarctica, an-*tark*-ti-ka. Of
Antarctic regions. Woolly Tree Fern.
fibrosa, fi-*bro*-sa. Fibrous (trunk).
Golden Tree Fern.
squarrosa, skwa-ro-sa. With the
parts spreading.

Dieffenbachia, dee-fan-*bark*-ee-a.
Araceae. After J. F. Dieffenbach.
Tender perennials. Dumb Cane.
amoena, a-*mee*-na. Pleasant.
bowmannii, bow-*man*-ee-ee. After
David Bowman.
exotica, ex-*o*-ti-ka. Exotic.
imperialis, im-peer-ee-*a*-lis. Showy.
maculata, mak-ew-*la*-ta. Spotted
(leaves).
oerstedi, ur-*sted*-ee-ee. After Anders
Oersted.

Dierama, dee-e-*ra*-ma. *Iridaceae.*
From Gk. *dierama* (funnel), the shape
of the flowers. Evergreen cormous
perennial.
pulcherrimum, pul-*ke*-ri-mum. Very
pretty. Angel's Fishing Rod, Wand
Flower.

Diervilla, dee-er-*vil*-la.
Caprifoliaceae. After Dr N. Dierville.
Deciduous shrub.
lonicera, lon-i-*se*-ra. *Lonicera-like.*

Digitalis, di-ji-*ta*-lis.
Scrophulariaceae. From L. *digitus*
(finger), after the finger-like flowers.
Biennial and perennial herbs.
  ambigua, am-*big*-ew-a. Doubtful.
  canariensis, ka-na-ree-*en*-sis. Of the
  Canary Islands.
  dubia, dub-ee-a. Doubtful.
  ferruginea, fe-roo-*jin*-ee-a. Rusty.
  Rusty Foxglove.
  grandiflora, gran-di-*flo*-ra. Large-
  flowered.
  lanata, la-*na*-ta. Woolly. Grecian
  Foxglove.
  lutea, loo-tee-a. Yellow.
  parviflora, par-vi-*flo*-ra. Small-flowered.
  purpurea, pur-*pur*-ree-a. Purple.
  Common Foxglove.

Digitalis purpurea

Dimorphotheca, di-mor-fo-*thee*-ka.
Compositae. From Gk. *dis* (twice),
morphe (shape) and *theka* (fruit), after
the shape of the fruit. Tender annual
and perennial herbs and evergreen sub-
shrubs.

  pluvialis, ploo-vee-*a*-lis. Of rain.
  sinuata, sin-ew-*a*-ta. Wavy-edged
  (leaves). Star of the Veldt.

Dionaea, die-on-*ee*-a. *Droseraceae.*
Gk. name for Venus. Evergreen, insectivorous perennial herbs.
  muscipula, mus-*kip*-ew-la. Venus's
  Fly Trap.

Dioscorea, die-os-*ko*-ree-a.
Dioscoreaceae. After Dioscorides.
Tender, perennial herbs and climbers.
Yam.
  discolor, dis-ko-lor. Two-coloured
  (leaves). Ornamental Yam.
  elephantipes, e-le-*fan*-ti-pees. Like
  an elephant's foot. Elephant's Foot,
  Tortoise Plant.

Diospyros, die-*os*-pi-ros. *Ebenaceae.*
From Gk. *dios* (divine) and *pyros*
(wheat), after the edible fruit.
Deciduous or evergreen trees and
shrubs.
  kaki, ka-ki. The Japanese name.
  Chinese Persimmon, Kaki.
  lotus, lo-tus. Date Plum.
  virginiana, vir-jin-ee-*a*-na. Of
  Virginia. Persimmon.

Dipelta, die-*pel*-ta. *Caprifoliaceae.*
From Gk. *di* (two) and *pelta* (shield),
the bracts enclosing the fruit.
Deciduous shrubs.
  floribunda, flo-ri-*bun*-da. Profusely
  flowering.
  ventricosa, ven-tri-*ko*-sa. Swollen.
  yunnanensis, yoo-nan-*en*-sis. Of
  Yunnan.

Dipsacus, *dip*-sa-kus, *Dipsacaceae.*
From Gk. *dipsa* (thirst). Water collects
in the leaf base cavities. Biennial or
perennial herb.
  fullonum, fu-*lo*-num. Of fullers.

Dipsacus fullonum

Common Teasel.
sativus, sa-*tee*-vus. Cultivated.
Fuller's Teasel.

Disanthus, dis-*anth*-us.
Hamamelidaceae. From Gk. *dis*
(twice) and *anthos* (flower), the flowers are in pairs. Deciduous shrub.
   cercidifolius, ser-si-di-*fo*-lee-us.
   Cercis-leaved.

Dizygotheca, di-zi-go-*thee*-ka.
Araliaceae. From Gk. *dis* (twice),
zygos (yoke) and *theka* (case). The
anthers have four lobes. Tender, evergreen trees and shrubs.
   elegantissima, e-le-gan-*tis*-i-ma.
   Most elegant. False Aralia.

Dodecatheon, do-dek-a-*thee*-on.
Primulaceae. From Gk. *dodeka*
(twelve) and *thios* (god). Perennial
herbs. Shooting Star.
   frigidum, fri-ji-dum. Of cold regions.
   hendersonii, hen-der-*son*-ee-ee.
   After Louis Fourniquet Henderson.

jeffreyi, jef-ree-ee. After Jeffrey.
meadia, mee-dee-a. After Richard
Mead.

Dombeya, *dom*-bee-a. *Sterculiaceae.*
After Joseph Dombey. Tender, evergreen trees and shrubs.
   burgessiae, bur-*jes*-ee-ee. After
   Miss Burgess.
   cayeuxii, kay-*yurz*-ee-ee. After
   Henri Cayeux.

Doronicum, do-*ron*-i-kum.
Compositae. Perennial herbs.
Leopard's Bane.
   austriacum, aw-stree-*a*-kum. Of
   Austria.
   columnae, ko-*lum*-nee. After Fabius
   Columna.
   plantagineum, plan-ta-*jin*-ee-um.
   Plantago-like.

Dorycnium, do-*rik*-nee-um.
Leguminosae. From Gk. *doryknion*
(*Convolvulus*). Perennial herbs and
deciduous or semi-evergreen sub-
shrubs.
   hirsutum, hir-*soo*-tum. Hairy.
   pedata, pe-*da*-ta. Like a bird's foot

Draba, *dra*-ba. *Cruciferae.* From Gk.
drabe, (cress). Annual and perennial
herbs.
   alpina, al-*pie*-na. Alpine.
   bruniifolia, brun-ee-i-*fo*-lee-a. With
   Brunia-like leaves.
   dedeana, dee-dee-*a*-na. After Dede.
   hispanica, hi-*spa*-ni-ka. Of Spain.
   lasiocarpa, la-see-o-*kar*-pa. Woolly-
   fruited.
   mollisima, mol-*lis*-i-ma. Very soft.
   polytricha, po-*li*-tri-ka. With many
   hairs.
   rigida, ri-ji-da. Rigid (leaves).
   imbricata, im-bri-*ka*-ta. Overlapping
   (leaves).

sibirica, si-*bi*-ri-ka. Of Siberia.

Dracaena, dra-*see*-na. *Agavaceae.*
From Gk. *drakaina* (dragon). Tender,
evergreen shrubs and trees.
  bausei, bowz-ee-ee. After Bause.
  deremensis, de-rem-*en*-sis. Of
  Derema, Tanzania.
  draco, dra-ko. A dragon. Dragon tree.
  fragrans, fra-granz. Fragrant (flowers).
  goldieana, gold-ee-*a*-na. After the
  Rev. Hugh Goldie.
  hookeriana, huk-a-ree-*a*-na. After
  Hooker.
  indivisa, in-di-*vee*-sa. Undivided.
  lindenii, lin-*den*-ee-ee. After Linden.
  marginata, mar-ji-*na*-ta. Margined
  (leaves).
  sanderiana, san-da-ree-*a*-na. After
  Henry Sander,
  surculosa, sur-kew-*lo*-sa. Suckering.

Dracocephalum, dra-ko-*sef*-a-lum.
Labiatae. From Gk. *draco* (dragon)
and *cephale* (head), after the flower
shape. Annual and perennial herbs.
  forrestii, fo-*rest*-ee-ee. After George
  Forrest.
  grandiflorum, gran-di-*flo*-rum.
  Large-flowered.
  hemsleyanum, hemz-lee-*a*-num.
  After William Botting Hemsley.

Dracunculus, dra-*kun*-kew-lus.
Araceae. L. for a small dragon.
Tuberous, perennial herbs.
  canariensis, ka-na-ree-*en*-sis. Of the
  Canary Islands.
  muscivorus, musk-*i*-vo-rus. Fly-eat-
  ing.
  vulgaris, vul-*ga*-ris. Common.
  Dragon Arum.

Drimys, *drim*-is. *Winteraceae.* From
the Gk. for acrid, the taste of the bark.

Evergreen trees and shrubs.
  lanceolata, lan-see-o-*la*-ta. Spear-
  shaped. Mountain Pepper.
  winteri, win-ta-ree. After Captain
  William Winter. Winter's Bark.

Drosera, *dro*-se-ra. *Droseraceae.*
From Gk. *droseros* (dewy). Evergreen,
insectivorous perennial herbs. Sundew.
  binata, bi-*na*-ta. Paired.
  capensis, ka-*pen*-sis. Of the Cape of
  Good Hope.
  filiformis, fee-lee-*form*-is. Thread-
  like (leaves).

Dryas, *drie*-as. *Rosaceae.* Gk. *dryas,*
from the Dryades, nymphs of the oak,
the leaves resemble the oak. Evergreen
shrubs.
  drummondii, dru-*mond*-ee-ee. After
  Thomas Drummond.
  octopetala, ok-to-*pe*-ta-la. Eight-
  petalled. Mountain Avens.

Dryas octopetala

Dryopteris, dree-*op*-te-ris.
Aspidiaceae. From Gk. *drys* (oak) and
pteris (fern). Deciduous or semi-evergreen ferns. Buckler Fern.
  austriaca, aw-stree-*a*-ka. Austrian.

dilatata, dil-a-*ta*-ta. Expanded (fronds). Broad Buckler Fern.

erythrosora, e-rith-ro-*so*-ra. With red sori.

filix-mas, fil-ix-mas. Male fern.

goldiana, gold-ee-*a*-na. After John Goldie. Giant Wood Fern.

Duchesnea, dew-*shez*-nee-a. *Rosaceae.* After Antoine Nicolas Duchesne. Perennial herbs.

indica, in-di-ka. Of India. Mock Strawberry.

chrysantha, kris-*anth*-a. With golden flowers.

E

Eccremocarpus, e-krem-o-*kar*-pus. *Bignoniaceae.* From Gk. *ekkremus* (hanging) and *karpos* (fruit), after the hanging pods. Evergreen, semi-hardy climbers.

    scaber, ska-ber. Rough.

Echeveria, e-kee-*ve*-ree-a. *Crassulaceae.* After Athanasio Echeverriay Godoy. Tender, perennial succulents.

    agavoides, a-gav-*oi*-deez. *Agave*-like.
    derenbergii, de-ran-*berg*-ee-ee. After J. Derenberg.
    elegans, e-le-ganz. Elegant.
    gibbiflora, jib-bi-*flo*-ra. With the flower swollen on one side.
    harmsii, harmz-ee-ee. After Dr Hermann Harms.
    multicaulis, mul-ti-*kaw*-lis. Many-stemmed.
    pulvinata, pul-vee-*na*-ta. Cushion-like.
    runyonii, run-*yon*-ee-ee. After Robert Runyon.
    secunda, se-*kun*-da. With flowers on one side of the stalk.
    glauca, glaw-ka. Smooth (leaves).
    setosa, see-*to*-sa. Bristly (stems).

Echinacea, e-kee-*na*-see-a. *Compositae.* From Gk. *echinos* (hedgehog), after the prickly scales. Perennial herb.

    purpurea, pur-*pur*-ree-a. Purple. (flowers).

Echinocactus, e-keen-o-*kak*-tus. *Cactaceae.* From Gk. *echinos* (hedgehog) and *Cactus.*

    grusonii, gru-*son*-ee-ee. After Herman Gruson. Golden Barrel Cactus.
    ingens, in-jens. Enormous.

Echinocereus, e-keen-o-*see*-ree-us. *Cactaceae.* From Gk. *echinos* (hedgehog) and *Cereus.*

    engelmannii, eng-gel-*man*-ee-ee. After Georg Engelmann.
    enneacanthus, en-ee-a-*kanth*-us. With nine spines.
    stramineus, stra-*min*-ee-us. Straw-coloured (spines).
    pectinatus, pek-tin-*a*-tus. Comb-like.
    rigidissimus, ri-ji-*dis*-i-mus. Very rigid.
    pentalophus, pen-ta-*lof*-us. With five crests.
    pulchellus, pul-*kel*-us. Pretty.
    viridiflorus, vi-ri-di-*flo*-rus. With green flowers.

Echinops, e-*kee*-nops. *Compositae.* From Gk. *echinos* (hedgehog) and *ops* (appearance). Perennial herbs. Globe Thistle.

    bannaticus, ba-*na*-ti-kus. Of Banat, Romania.
    humilis, hu-mi-lis. Low growing.
    sphaerocephalus, sfee-ro-*sef*-a-lus. With a round head.

Echinopsis, e-kee-*nop*-sis. *Cactaceae.* From Gk. *echinos* (hedgehog) and *-opsis* (appearance). Sea Urchin Cactus.

    aurea, aw-ree-a. Golden.
    eyriesii, ie-*reez*-ee-ee. After Alexander Eyries.
    ferox, fe-rox. Spiny.

leucantha, loo-*kanth*-a. White-flow-
ered.
multiplex, mul-ti-plex. With many
stems.
rhodotricha, ro-*do*-tri-ka. Red-
haired.

Echium, e-*kee*-um. *Boraginaceae.*
From Gk. *echion* (viper). Biennial
herbs.
   plantagineum, plan-ta-*jin*-ee-um.
   Plantago-like .
   russicum, ru-si-kum. Russian
   vulgare, vul-*ga*-ree. Common.
   Viper's Bugloss.

Echium plantagineum

Edgeworthia, ej-*werth*-ee-a.
Thymelaeaceae. After Michael
Pakenham Edgeworth. Semi-hardy,
deciduous shrub.
   chrysantha, kris-*anth*-a. With golden
   flowers.
   papyrifera, pa-pi-*ri*-fe-ra. Paper-
   bearing.

Edraianthus, ed-rie-*anth*-us.
Campanulaceae. From Gk. *hedraios*
(sitting) and *anthos* (flower) after the
stalkless flowers. Perennial herbs.
   dalmaticus, dal-*ma*-ti-kus. Of
   Dalmatia.
   pumilio, pew-*mil*-lee-o. Dwarf.
   serpyllifolius, ser-pi-li-*fo*-lee-us.
   Thyme-leaved.

Eichhornia, iek-*horn*-ee-a.
Pontederiaceae. After J. A. F.
Eichhorn. Evergreen or semi-evergreen, perennial aquatic herbs.
   crassipes, kras-i-pees. With a thick
   stalk (leaves). Water Hyacinth.

Elaeagnus, e-lee-*ag*-nus.
Elaeagnaceae. From Gk. *helodes*
(marsh-growing) and *hagnos* (pure).
Deciduous or evergreen trees and shrubs.
   angustifolia, an-gust-i-*fo*-lee-a.
   Narrow-leaved.
   commutata, kom-ew-*ta*-ta.
   Changeable. Silver Berry.
   glabra, gla-bra. Smooth.
   macrophylla, mak-ro-*fil*-a. Large-
   leaved.
   maculata, mak-ew-*la*-ta. Blotched
   (leaves).
   pungens, pun-jenz. Sharp-pointed.
   umbellata, um-bel-*a*-ta. With flowers in umbels.

Eleocharis, e-lee-*o*-ka-ris.
Cyperaceae. From Gk. *helodes*
(marsh-growing) and *charis* (grace).
Evergreen, perennial aquatic herbs.
   acicularis, a-sik-ew-*la*-ris. Needle-
   like (stems). Hair Grass.
   dulcis, dul-sis. Sweet. Chinese Water
   Chestnut.

Elsholtzia, el-*sholtz*-ee-a. *Labiatae.*
After Johann Sigismund Elsholtz.
Deciduous perennial shrubs and sub-
shrubs.
   stauntonii, stawn-*ton*-ee-ee. After Sir
   George Staunton.

Embothrium, em-*both*-ree-um.
Proteaceae. From Gk. *en* (in) and
bothrion (small pit). Evergreen or
semi-evergreen trees.
    coccineum, kok-*kin*-ee-um. Scarlet.
Fire Bush.

Emilia, em-*ee*-lee-a. *Compositae.*
Semi-hardy, annual and perennial
herbs.
    javanica, ja-*va*-ni-ka. Of Java.
Tassel Flower.

Emmenopterys, e-men-*op*-te-ris.
Rubiaceae. From Gk. *emmenes* (endur-
ing) and *pteryx* (wing). Deciduous
trees.
    henryi, hen-ree-ee. After Augustine
Henry.

Empetrum, *em*-pe-trum.
Empetraceae. From Gk. *en* (on) and
petros (rock), after its habitat.
Evergreen shrub. Crowberry.
    nigrum, *nig*-rum. Black (fruit).

Enkianthus, eng-kee-*anth*-us.
Ericaceae. From Gk. *enknos* (preg-
nant) and *anthos* (flower). Deciduous

Empetrum nigrum

or semi-evergreen trees and shrubs.
    campanulatus, kam-pan-ew-*la*-tus.
Bell-shaped (corolla).
    cernuus, *ser*-new-us. Nodding
(racemes).
    rubens, *roo*-benz. Red (flowers).
    chinensis, chin-*en*-sis. Of China

Epidendrum, e-pi-*den*-drum.
Orchidaceae. From Gk. *epi* (upon) and
dendron (tree), growing in trees.
Greenhouse orchids.
    ciliare, si-lee-*a*-ree. Edged with
hairs (lip).
    cochleatum, kok-lee-*a*-tum. Shell-
like (lip).
    difforme, di-*form*-ee. Of unusual
shape.
    fragrans, *fra*-granz. Fragrant.
    mariae, *ma*-ree-ee. After Mrs Mary
Ostlund.
    nocturnum, nok-*tur*-num. Night
flowering.
    parkinsonianum, par-kin-son-ee-*a*-
num. After John Parkinson.
    polybulbon, pol-i-*bul*-bon. With
many bulbs.
    radiatum, ra-dee-*a*-tum. Radiating.

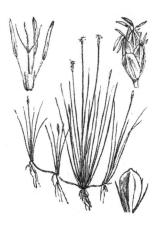

Eleocharis acicularis

radicans, ra-di-kanz. With rooting stems.
vitellinum, vi-te-*leen*-um. Colour of Egg-yolk.

Epigaea, e-pi-*jee*-a. *Ericaceae.* From Gk. *epi* (on) and *gaia* (earth). Evergreen sub-shrubs.
asiatica, a-see-*a*-ti-ka. Of Asia.
repens, ree-penz. Creeping. Trailing Arbutus.

Epilobium, e-pi-*lo*-bee-um. *Onagraceae.* From Gk. *epi* (upon) and *lobos* (pod). Annual and perennial herbs and deciduous sub-shrubs. Willow Herb.
album, al-bum. White.
angustifolium, an-gust-i-*fo*-lee-um. Narrow-leaved.
chlorifolium, klo-ri-*fo*-lee-um. With *Chlora*-like leaves.
dodonaei, do-do-*nee*-ee. After Rembert Dodoens.
fleischeri, flie-sha-ree. After M. Fleischer.
glabellum, gla-*bel*-um. Almost smooth.
latifolium, la-ti-*fo*-lee-um. Broad-leaved.

Epimedium, e-pi-*mee*-dee-um. *Berberidaceae.* From Gk. *epimedion.* Perennial herbs. Bishop's Hat.
alpinum, al-*pie*-num. Alpine.
grandiflorum, grand-i-*flo*-rum. Large-flowered.
perralderianum, pe-ral-de-ree-*a*-num. After Henri Rene le Tourneux de la Perraudiere.
pinnatum, pin-*a*-tum. Pinnate.
pubigerum, pew-*bi*-je-rum. Hairy.
rubrum, rub-rum. Red.
setosum, see-*to*-sum. Bristly.
versicolor, ver-*si*-ko-lor. Variously coloured.

Epiphyllum, e-pi-*fil*-lum. *Cactaceae.* From Gk. *epi* (upon) and *phyllon* (leaf). Orchid Cactus.
ackermannii, a-ker-*man*-ee-ee. After Georg Ackermann.
anguliger, ang-*gew*-li-jer. Hooked.
caudatum, kaw-*da*-tum. With a tail (shoots).
chrysocardium, kris-o-*kar*-dee-um. With a golden heart.
crenatum, kree-*na*-tum. Crenated.

Epipremnum, e-pi-*prem*-num, *Araceae.* From Gk. *epi* (upon) and *premnum* (tree stump), growing in trees. Tender, evergreen climbers.
aureum, aw-ree-um. Golden. Devil's Ivy.
pictum, pik-tum. Painted.

Episcia, e-*pis*-ee-a, *Gesneriaceae.* From Gk. *episkios* (shaded), growing in shady places. Tender, evergreen perennial herbs.
cupreata, kew-pree-*a*-ta. Copper (leaves). Flame Violet.
dianthiflora, dee-anth-i-*flo*-ra. *Dianthus*-flowered. Lace Flower Vine.
lilacina, li-la-*seen*-a. Lilac.
metallica, me-*ta*-li-ka. Metallic.
reptans, rep-tanz. Creeping.

Eragrostis, e-ra-*gros*-tis. *Gramineae.* From Gk. *eros* (love) and *agrostis* (a grass). Annual grasses.
amabilis, a-*ma*-bi-lis. Beautiful. Japanese Love Grass.
elegans, e-le-ganz. Elegant.

Eranthis, e-*ran*-this. *Ranunculaceae.* From Gk. *er* (spring) and *anthos* (flower), after the early flowers. Perennial, tuberous herbs.
cilicica, si-*liss*-see-a. Of Cilicia, Turkey.

hyemalis, hie-e-*ma*-lis. Of winter. Winter Aconite.

Eremurus, e-ree-*mew*-rus. *Liliaceae.* From Gk. *eremia* (desert) and *oura* (tail). Perennial herbs. Foxtail Lily.
  elwesii, el-*wez*-ee-ee. After H. J. Elwes.
  himalaicus, hi-ma-*la*-i-kus. Of the Himalayas.
  olgae, ol-gee. After Olga Fedtschenko.
  robustus, ro-*bust*-us. Robust.
  spectabilis, spek-*ta*-bi-lis. Spectacular.
  stenophyllus, sten-o-*fil*-lus. Narrow-leaved.

Erica, e-*ree*-ka. *Ericaceae.* Tender and hardy, evergreen trees and sub-shrubs. Heath.
  arborea, ar-*bo*-ree-a. Tree-like. Tree Heath.
  alpina, al-*pie*-na. Alpine.
  australis, aw-*stra*-lis. Southern. Spanish Heath.
  canaliculata, kan-a-lik-ew-*la*-ta. Channelled.

Erica cinerea

carnea, kar-nee-a. Flesh-coloured.
cinerea, si-*ne*-ree-a. Grey. Bell Heather.
erigena, e-ri-*gen*-a. Irish.
gracilis, *gra*-si-lis. Graceful.
herbacea, her-*ba*-see-a. Herbaceous.
hyemalis, hie-e-*ma*-lis. Of winter.
lusitanica, loo-si-*ta*-ni-ka. From Portugal.
terminalis, ter-mi-*na*-lis. Terminal (flowers).
vagans, va-ganz. Wandering.
ventricosa, ven-tri-*ko*-sa. Swollen on one side.

Erigeron, e-*ri*-je-ron. *Compositae.* From Gk. *eri* (early) and *geron* (old man), after the white seed heads. Annual and Perennial herbs. Fleabane.
  alpinus, al-*pie*-nus. Alpine.
  aurantiacus, aw-ran-tee-*a*-kus. Orange.
  aureus, aw-ree-us. Golden.
  compositus, kom-*po*-si-tus. Compound.
  glaucus, *glaw*-kus. Glaucous. Beach Aster.
  karvinskianus, kar-vin-skee-*a*-nus. After Wilhelm Friedrich Karwinski von Karwin.
  simplex, *sim*-plex. Simple.
  speciosus, spes-ee-*o*-sus. Showy.
  macranthus, ma-*kranth*-us. Large-flowered.

Erinacea, e-ri-*na*-see-a. *Leguminosae.* L. for resembling a hedgehog. Evergreen, spiny sub-shrubs.
  anthyllis, an-*thil*-lis. Kidney vetch. Hedgehog Broom.
  pungens, *pun*-jenz. Sharp-pointed.

Erinus, *e*-ri-nus. *Scrophulariaceae.* Semi-evergreen, perennial herbs.
  alpinus, al-*pie*-nus. Alpine. Fairy Foxglove.

Eriobotrya, e-ree-o-*bot*-ree-a.
Rosaceae. From Gk. *erion* (wool) and
botrys (bunch of grapes). Evergreen,
semi-hardy trees and shrubs.
  japonica, ja-*pon*-i-ka. Of Japan.
  Loquat.

Eritrichium, e-ri-*trik*-ee-um.
Boraginaceae. From Gk. *erion* (wool)
and *thrix* (hair). Perennial herbs.
  canum, ka-num. Grey.
  nanum, na-num. Dwarf. Fairy
  Forget-me-not, King of the Alps.
  rupestre, roo-*pes*-tree. Growing on
  rocks.

Erodium, e-*ro*-dee-um. *Geraniaceae.*
From Gk. *erodios* (heron). Perennial
herbs. Heron's Bill.
  absinthoides, ab-sinth-*oi*-deez. Like
  Artemisia absinthium.
  chrysanthum, kris-*anth*-um. With
  golden flowers.
  corsicum, kor-si-kum. Of Corsica.
  guttatum, gu-*ta*-tum. Spotted.
  petraeum, pe-*tree*-um. Growing in
  rocky places.
  glandulosum, glan-dew-*lo*-sum.
  Glandular.
  reichardii, rie-*kard*-ee-ee. After
  Reichard.
  roseum, ro-see-um. Rose-coloured.

Eryngium, e-*rin*-jee-um.
Umbelliferae. From Gk. *eryggion.*
Biennial and perennial herbs.
  alpinum, al-*pie*-num. Alpine.
  amethystinum, a-me-*thist*-i-num.
  Violet.
  bourgatii, bour-*gat*-ee-ee. After M.
  Bourgat.
  giganteum, ji-*gan*-tee-um. Very
  large.
  maritimum, ma-*ri*-ti-mum. Growing
  near the sea. Sea Holly.
  planum, pla-num. Flat.

Eryngium maritimum

  tripartitum, tri-*part*-ee-tum. In three
  parts.

Erysimum, e-*ri*-si-mum. *Cruciferae.*
From Gk. *erysimon.* Semi-evergreen,
annual, biennial and perennial herbs.
  asperum, a-*spe*-rum. Rough (leaves).
  capitatum, kap-i-*ta*-tum. In a dense
  head (flowers).
  helveticum, hel-*vee*-ti-kum. Of
  Switzerland.
  linifolium, lin-i-*fo*-lee-um. *Linum*-
  leaved.
  murale, mew-*ra*-lee. Growing on
  walls.
  pulchellum, pul-*kel*-um. Pretty.

Erythrina, e-rith-*reen*-a.
Leguminosae. From Gk. *erythros* (red),
after the colour of the flowers. Tender,
deciduous or semi-evergreen trees and
shrubs.
  crista-galli, kris-ta-*ga*-lee. Cock's
  comb. Coral Tree.

Erythronium, e-rith-*ron*-ee-um.
Liliaceae. From Gk. *erythronion.*
Tuberous, perennial herbs.

albidum, al-bi-dum. White.

americanum, a-me-ri-*ka*-num. Of America. Yellow Adder's Tongue.

californicum, kal-i-*forn*-i-kum. Of California.

citrinum, si-*tree*-num. Lemon-yellow (flowers).

dens-canis, dens-*ka*-nis. Dog's tooth. Dog's-tooth Violet.

grandiflorum, grand-i-*flo*-rum. Large-flowered.

howellii, how-*el*-ee-ee. After Thomas Howell.

multiscapoideum, mul-ti-ska-*poi*-dee-um. With many scapes.

oregonum, o-ree-*go*-num. Of Oregon.

revolutum, re-vo-*loo*-tum. Turned back. Trout Lily.

Escallonia, es-ka-*lon*-ee-a. Crassulariaceae. After Senor Escallon. Evergreen or deciduous trees and shrubs.

bifida, bi-fi-da. Split into two.

edinensis, e-din-*en*-sis. Of Edinburgh

iveyi, ie-vee-ee. After Mr Ivey.

laevis, lee-vis. Smooth (leaves).

rubra, ru-bra. Red (flowers).

macrantha, ma-*kranth*-a. Large-flowered.

virgata, vir-*ga*-ta. Twiggy.

Eschscholzia, esh-*sholts*-ee-a. *Papaveraceae.* After Johann Friedrich Eschscholz. Annual herbs.

caespitosa, see-spi-*to*-sa. Tufted.

californica, kal-i-*forn*-i-ka. Of California. Californian Poppy.

Eucalyptus, ew-ka-*lip*-tus. *Myrtaceae.* From Gk. *eu* (well) and *kalypto* (cover). Hardy to tender evergreen trees. Gum Tree.

citriodora, sit-ree-o-*do*-ra. Lemon-scented. Lemon-scented Gum

coccifera, kok-*kif*-e-ra. Berry-bearing. Mount Wellington Peppermint, Tasmanian Snow Gum.

cordata, kor-*da*-ta. With heart-shaped leaves. Silver Gum.

dalrympleana, dal-rim-plee-*a*-na. After Dalrymple.

ficifolia, fi-ki-*fo*-lee-a. *Ficus*-leaved.

globulus, glob-ew-lus. Like a small globe (buds). Tasmanian Blue Gum.

gunnii, gun-ee-ee. After Ronald Gunn.

parvifolia, par-vi-*fo*-lee-a. Small-leaved. Small-leaved Gum.

pauciflora, paw-si-*flo*-ra. With few flowers.

urnigera, ur-*ni*-ge-ra. Urn-bearing. Urn Gum.

viminalis, vee-min-*a*-lis. With long slender shoots.

Eucomis, ew-*kom*-is. *Liliaceae.* From Gk. *eu* (good) and *kome* (hair). Bulbous herbs. Pineapple Flower.

autumnalis, aw-tum-*na*-lis. Of autumn.

bicolor, bi-ko-lor. Two-coloured (flowers).

comosa, ko-*mo*-sa. With a tuft.

pallidiflora, pa-li-di-*flo*-ra. Pale-flowered.

Eucommia, ew-*kom*-ee-a. *Eucommiaceae.* From Gk. *eu* (good) and *kommi* (gum). Deciduous tree.

ulmoides, ul-*moi*-deez. Like *Ulmus.* Gutta-percha Tree.

Eucryphia, ew-*krif*-ee-a. *Eucryphiaceae.* From Gk. *eu* (well) and *kryphios* (covered). Hardy and semi-hardy, semi-evergreen trees and shrubs.

cordifolia, kor-di-*fo*-lee-a. With heart-shaped leaves.

glutinosa, gloo-ti-*no*-sa. Sticky.
lucida, loo-si-da. Glossy.
milliganii, mil-li-*gan*-ee-ee. After
Joseph Milligan.

Euodia, ew-*o*-dee-a. *Rutaceae.* From
Gk. *euodia* (sweet scent), fragrant
flowers. Deciduous trees.
hupehensis, hew-pee-*hen*-sis. Of
Hubeh, China.

Euonymus, ew-*on*-i-mus.
Celastraceae. Deciduous or evergreen
trees and shrubs.
alatus, a-*la*-tus. Winged (shoots).
europaeus, ew-ro-*pee*-us. European.
Spindle Tree.
fortunei, for-*tewn*-ee-ee. After
Robert Fortune.
grandiflorus, grand-i-*flo*-rus. Large-
flowered.
hamiltonianus, ha-mil-ton-ee-*a*-nus.
After Francis Buchanan-Hamilton.
japonicus, ja-*pon*-i-kus. Of Japan.
latifolius, la-ti-*fo*-lee-us. Broad-
leaved.
nanus, na-nus. Dwarf.
oxyphyllus, ox-i-*fil*-lus. With sharp-
pointed leaves.
planipes, plan-i-pees. With a flat
stalk.
wilsonii, wil-*son*-ee-ee. After Ernest
Wilson.

Eupatorium, ew-pa-*to*-ree-um.
Compositae. After Eupator, King of
Pontus. Tender, perennial herbs, shrubs
and sub-shrubs.
cannabinum, kan-a-*been*-um.
Cannabis-like. Hemp Agrimony.
coelestinum, see-les-*teen*-um. Sky-
blue. Mistflower.
ligustrinum, lig-us-*tree*-num. Like
Ligustrum.
maculatum, mak-ew-*la*-tum.
Spotted. Joe-pye Weed.

Eupatorium cannabinum

purpureum, pur-*pur*-ree-um. Purple.
rugosum, roo-*go*-sum. Wrinkled.

Euphorbia, ew-*for*-bee-a.
Euphorbiaceae. After Euphorbus.
Tender to hardy, annual and perennial
shrubs, sub-shrubs and succulents.
Milkweed.
candelabrum, kan-dee-*la*-brum. a
Candelabra-like.
caput-medusae, ka-put-mee-*dew*-see.
Medusa's head.
cyathophora, sie-ath-o-*fo*-ra. Cup-
bearing. Fire on the Mountain.
cyparissias, si-pa-*ris*-ee-as. Cypress-
like. Cypress Spurge.
echinus, e-*keen*-us. Spiny.
fulgens, ful-jenz. Shining (bracts).
grandicornis, grand-i-*kor*-nis. With
large horns.
griffithii, gri-*fith*-ee-ee. After
Griffith.
mammillaris, ma-mi-*la*-ris. Bearing
nipples. Corncob Cactus.
marginata, mar-ji-*na*-ta. Margined
(leaves). Snow on the Mountain.
meloformis, me-lo-*form*-is. Melon-
shaped.

myrsinites, mur-sin-*ee*-teez. *Myrsine*-like.

niciciana, nee-cheech-ee-*a*-na. After Nicic.

obesa, o-bee-sa. Fat. Baseball Cactus.

palustris, pa-*lus*-tris. Growing in marshes.

pseudocactus, soo-do-*kak*-tus. False cactus.

pulcherrima, pul-*ke*-ri-ma. Very pretty. Poinsettia.

resinifera, re-see-*ni*-fe-ra. Resin-bearing.

rigida, ri-ji-da. Rigid.

sikkimensis, sik-im-*en*-sis. Of Sikkim.

splendens, splen-denz. Splendid

submammillaris, sub-ma-mi-*la*-ris. With small nipples.

valida, va-li-da. Robust.

Eurya, *ew*-ree-a. *Theaceae.* From Gk. *euru* (broad). Evergreen trees and shrubs.

chinensis, chin-*en*-sis. Of China.

japonica, ja-*pon*-i-ka. Of Japan.

Euryops, *ew*-ree-ops. *Compositae.* From Gk. *eu* (well) and *ops* (appearance). Evergreen shrubs and sub-shrubs.

acraeus, a-*kree*-us. Growing in high places.

pectinatus, pek-ti-*na*-tus. Comb-like.

Exochorda, ex-o-*kor*-da. *Rosaceae.* From Gk. *exo* (outside) and *chorda* (cord). Deciduous shrubs.

giraldii, ji-*ral*-dee-ee. After Giraldi.

korolkowii, ko-rol-*kov*-ee-ee. After Korolkow.

macrantha, ma-*kranth*-a. Large-flowered.

racemosa, ra-see-*mo*-sa. With flowers in racemes.

Exacum, *ex*-a-kum. *Gentianaceae.* Tender annual or biennial.

affine, a-*fee*-nee. Related to. Persian Violet.

F

Fabiana, fa-bee-*a*-na. *Solanaceae.*
After Archbishop Francisco Fabian y
Fuero. Evergreen, semi-hardy shrub.
  imbricata, im-bri-*ka*-ta. Closely
  overlapping (leaves).

Fagus, *fay*-gus. *Fagaceae.* Deciduous
trees. Beech.
  americana, a-me-ri-*ka*-na. Of
  America
  grandifolia, grand-i-*fo*-lee-a. Large-
  leaved.
  orientalis, o-ree-en-*ta*-lis. Eastern.
  pendula, *pen*-dew-la. Pendulous.
  purpurea, pur-*pur*-ree-a. Purple
  (leaves). Copper Beech.
  sylvatica, sil-*va*-ti-ka. Of woods.
  Common Beech.

Fagus sylvatica

Fatsia, *fat*-see-a. *Araliaceae.* From a
Japanese name. Evergreen shrubs.
  japonica, ja-*pon*-i-ka. Of Japan.
  False Castor-oil Plant.

  papyrifera, pa-pi-*ri*-fe-ra. Papery.

Faucaria, fow-*ka*-ree-a. *Aizoaceae.*
From L. *faux* (gullet). The leaves
resemble open jaws. Tender perennial
succulents.
  tigrina, tig-*reen*-a. Tiger-like. Tiger's
  Jaws.
  tuberculosa, tew-ber-kew-*lo*-sa.
  Tubercled.

Feijoa, fie-*jo*-a. *Myrtaceae.* After Don
de Silva Feijo. Evergreen semi-hardy
shrubs.
  sellowiana, se-lo-ee-*a*-na. After
  Friedrich Sellow.

Felicia, fe-*liss*-ee-a. *Compositae.* After
Felix. Annuals and evergreen sub-
shrubs.
  amelloides, a-mel-*oi*-deez. Like
  Aster amellus.
  bergeriana, ber-ga-ree-*a*-na. After
  Berger. Kingfisher Daisy.
  rosulata, ros-ew-*la*-ta.With leaves in
  a rosette.

Ferocactus, fe-ro-*kak*-tus. *Cactaceae.*
From L. *ferox* (savage), after the
spines, and *Cactus.*
  acanthodes, a-*kanth*-o-deez. Spiny.
  hamatacanthus, ha-ma-ta-*kanth*-us.
  With hooked spines.
  latispinus, la-ti-*speen*-us. With broad
  spines.
  setispinus, se-tee-*speen*-us. With
  bristle-like spines.
  viridescens, vi-ri-*des*-enz. Greenish
  (flowers).
  wislizenii, wiz-li-*zen*-ee-ee. After
  Wislizenius.

Ferula, *fe*-ru-la. *Umbelliferae*. L.
name. Perennial herbs.
    communis, kom-*ew*-nis. Common.
Giant Fennel.
    tingitana, tin-ji-*ta*-na. Of Tangier.

Festuca, fes-*too*-ka. *Gramineae*. L. for
a grass stalk. Evergreen perennial
grasses.
    alpina, al-*pie*-na. Alpine.
    amethystina, a-me-*this*-ti-na. Violet.
    glacialis, gla-see-*a*-lis. Growing in
icy places.
    glauca, *glaw*-ka. Smooth (leaves).

Ficus, *fi*-kus. *Moraceae*. The L. for *F.
carica*. Deciduous or evergreen trees,
shrubs and climbers.
    benghalensis, beng-ga-*len*-sis. Of
Benghal. Banyan Tree.
    benjamina, ben-ja-*meen*-a. From
benjan, the Indian name. Weeping
Fig.
    carica, *ka*-ri-ka. Of Caria. Common
Fig.
    deltoidea, del-*toi*-dee-a. Triangular
(leaves). Mistletoe Fig.
    elastica, e-*las*-ti-ka. Elastic Rubber
Plant.
    lyrata, li-*ra*-ta. Fiddle-shaped
(leaves).
    macrophylla, mak-ro-*fil*-a. Large-
leaved.
    pumila, *pew*-mi-la. Dwarf. Creeping
Fig.
    religiosa, re-lij-ee-*o*-sa. Sacred.
    retusa, re-*tew*-sa. Notched at the
apex (leaves).
    rubiginosa, roo-bi-ji-*no*-sa. Rusty.
Rusty Fig.
    sagittata, saj-i-*ta*-ta. Shaped like an
arrow-head (leaves).

Filipendula, fi-li-*pen*-dew-la.
Rosaceae. From L. *filum* (thread) and
pendulus (hanging), threads connect

the root tubers. Perennial herbs.
    camtschatica, kamt-*sha*-ti-ka. Of
Kamtchatka.
    palmata, pal-*ma*-ta. Lobed like a
hand (leaves).
    purpurea, pur-*pur*-ree-a. Purple.
    rubra, *rub*-ra. Red. Queen of the
Prairie.
    ulmaria, ul-*ma*-ree-a. *Ulmus*-like
(leaflets). Meadowsweet.
    vulgaris, vul-*ga*-ris. Common.
Dropwort.

Fittonia, fi-*ton*-ee-a. *Acanthaceae*.
After Elizabeth and Sarah Mary Fitton.
Tender, evergreen, creeping perennial
herbs.
    argyroneura, ar-ji-ro-*new*-ra. Silver-
veined. Mosaic Plant.
    gigantea, ji-*gan*-tee-a. Very large.
    verschaffeltii, vair-sha-*felt*-ee-ee.
After M. Verschaffelt.

Fitzroya, fitz-*roy*-a. *Cupressaceae*.
After Captain Robert Fitzroy.
Evergreen conifer.
    cupressoides, kew-pres-*oi*-deez,

Foeniculum vulgare

Cupressus-like.
patagonica, pat-a-*gon*-i-ka. Of
Patagonia.

Foeniculum, fee-*nik*-ew-lum.
Umbelliferae. L. name. Biennial and
perennial herbs.
    azoricum, a-*zo*-ri-kum. Of the
Azores. Florence Fennel.
    vulgare, vul-*ga*-ree. Common.
Fennel.

Fontinalis, fon-ti-*na*-lis.
Fontinalaceae. From L. *fontinalis*
(springs or fountains). Evergreen,
perennial aquatic moss.
    antipyretica, an-ti-pi-*ret*-i-ka.
Against fire. Willow Moss.

Fragaria vesca

Forsythia, for-*sie*-thee-a. *Oleaceae.*
After William Forsyth. Deciduous
shrubs.
    ovata, o-*va*-ta. Ovate (leaves).
Korean Forsythia.
    suspensa, sus-*penz*-a. Hanging
(flowers). Golden Bell.
    viridissima, vi-ri-*di*-si-ma. Most
green (shoots).

Fortunella, for-tew-*nel*-a. *Rutaceae.*
After Robert Fortune.
    japonica, ja-*pon*-i-ka. Of Japan.
Kumquat.

Fothergilla, fo-tha-*gil*-a.
Hamamelidaceae. After Dr John
Fothergill. Deciduous shrubs.
    gardenii, gar-*den*-ee-ee. After Dr
Garden.
    major, ma-jor. Larger.
    monticola, mon-*ti*-ko-la. Mountain-
loving.

Fragaria, fra-*ga*-ree-a. *Rosaceae.*
From L. *fraga* (sweet-smelling).
Perennial herbs. Strawberry.

x *ananassa,* an-a-*nas*-a. From
Ananas. Garden Strawberry.
    indica, in-dik-a. Of India.
    moschata, mos-*ka*-ta. Musk-scented.
Hautbois Strawberry.
    vesca, ves-ka. Little. Wild
Strawberry.

Francoa, fran-*ko*-a. *Saxifragaceae.*
After Francisco Franco. Perennial
herbs.
    sonchifolia, son-ki-*fo*-lee-a. With
leaves like *Sonchus* (Sowthistle).
Bridal Wreath.

Frankenia, fran-*ken*-ee-a.
Frankeniaceae. After Johan
Frankenius. Evergreen sub-shrubs or
perennial herbs.
    thymifolia, tie-mi-*fo*-lee-a. *Thymus*-
leaved.
    laevis, lee-vis. Smooth.

Franklinia, frank-*lin*-ee-a. *Theaceae.*
After Benjamin Franklin. Deciduous
tree or shrub.
    alatamaha, a-la-ta-*ma*-ha. Of the
Altamaha River.

Fraxinus, *frax*-i-nus. *Oleaceae*. L. name. Deciduous trees and shrubs. Ash.

americana, a-me-ri-*ka*-na. Of America. White Ash.

angustifolia, ang-gus-ti-*fo*-lee-a. Narrow-leaved. Narrow-leaved Ash.

excelsior, ex-*sel*-see-or. Taller. Common Ash.

mariesii, ma-*reez*-ee-ee. After Maries.

ornus, or-nus. L. for the mountain ash. Manna Ash.

oxycarpa, ox-i-*kar*-pa. With pointed fruits.

velutina, vel-ew-*teen*-a. Velvety. Arizona Ash.

Fraxinus excelsior

Freesia, *freez*-ee-a. *Iridaceae*. After Friedrich Heinrich Theodor Freese. Tender or semi-hardy cormous perennials.

alba, al-ba. White (flowers).

corymbosua, ko-rim-*bo*-sua. With flowers in corymbs.

x *hybrida, hib*-ri-da. Hybrid. Common Freesia.

lactea, lak-tee-a. Milky

refracta, re-*frak*-ta. Broken.

Fremontodendron, free-mont-o-*den*-dron. *Sterculiaceae*. (*Fremontia*). After Major-General John Charles Fremont.

californicum, kal-i-*forn*-i-kum. Of California.

mexicanum, mex-i-*ka*-num. Mexican.

Fritillaria, fri-ti-*la*-ree-a. *Liliaceae*. From L. *fritillus* (dicebox), after the chequered flowers. Bulbous perennials. Fritillary.

acmopetala, ak-mo-*pe*-ta-la. With anvil-shaped petals.

bithynica, bi-*thin*-i-ka. Of Bithynia.

camschatsensis, kam-shat-*sen*-sis. Of Kamtchatka.

caucasica, kaw-*kas*-i-ka. Of the Caucasus.

cirrhosa, si-r*o*-sa. With tendrils.

crassifolia, kra-si-*fo*-lee-a. Thick-leaved.

davisii, day-*vis*-ee-ee. After Peter Hadland Davis.

graeca, gree-ka. Of Greece.

imperialis, im-peer-ee-*a*-lis. Showy. Crown Imperial.

Fritillaria meleagris

involucrata, in-vo-loo-*kra*-ta. With an involucre.

lanceolata, lan-see-o-*la*-ta. Spear-shaped (leaves).

latifolia, la-ti-*fo*-lee-a. Broad-leaved.

lusitanica, loo-si-*ta*-ni-ka. Of Portugal.

meleagris, mel-ee-*a*-gris. Spotted (flowers).

messanensis, mes-an-*en*-sis. Of Messina, Sicily.

pallidiflora, pa-li-di-*flo*-ra. Pale-flowered.

persica, per-si-ka. Of Persia (Iran).

pluriflora, ploo-ri-*flo*-ra. Many-flowered.

pontica, pon-ti-ka. Of Pontus.

pudica, pud-*ee*-ka. Modest.

pyrenaica, pi-ren-*ee*-i-ka. Of the Pyrenees.

raddeana, ra-dee-*a*-na. After Gustav Ferdinand Richard Radde.

recurva, re-*kur*-va. Curved back.

verticillata, ver-ti-sil-*a*-ta. Whorled (leaves).

Fuchsia, *few*-she-a. *Onagraceae.* After Leonhart Fuchs. Deciduous and ever-green, semi-hardy and tender shrubs.

excorticata, ex-sor-ti-*ka*-ta. With peeling bark.

gracilis, gra-si-lis. Graceful.

magellanica, ma-jel-*an*-i-ka. From the region of the Magellan Straits.

procumbens, pro-*kum*-benz. Prostrate. Trailing Fuchsia.

G

Gagea lutea

Gagea, *gay*-jee-a. *Liliaceae*. After Sir
Thomas Gage. Bulbous perennials.
   fistulosa. fist-u-*lo*-sa. With hollow
   leaves.
   graeca, gree-ka. Of Greece.
   lutea, loo-tee-a. Yellow (flowers).
   pratensis, pra-*ten*-sis. Of meadows.

Gaillardia, gay-*lard*-ee-a.
Compositae. After Gaillard de
Charentonneau. Annual and perennial
herbs. Blanket Flower.
   aristata, a-ris-*ta*-ta. Bearded.
   grandiflora, grand-i-*flo*-ra. Large-
   flowered.
   pulchella, pul-*kel*-a. Pretty.

Galanthus, ga-*lanth*-us.
Amaryllidaceae. From Gk. *gala* (milk)
and *anthos* (flower), after the colour of
the flowers. Bulbous herbs. Snowdrop.
   allenii, a-*len*-ee-ee. After James Allen.
   alpinus, al-*pie*-nus. Alpine.

byzantinus, bi-zan-*teen*-us. Of
Byzantium (Istanbul).
caucasicus, kaw-*ka*-si-cus. Of the
Caucasus.
elwesii, el-*wez*-ee-ee. After H. J.
Elwes.
fosteri, fos-ta-ree. After Sir Michael
Foster.
gracilis, gra-si-lis. Graceful.
ikariae, i-*ka*-ree-ee. Of Ikaria.
latifolius, la-ti-*fo*-lee-us. Broad-
leaved.
nivalis, ni-*va*-lis. Of the snow.
Common Snowdrop.
plicatus, pli-*ka*-tus. Pleated (leaves).

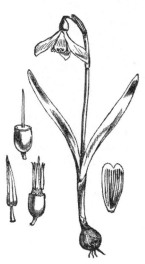

Galanthus nivalis

Galax, *gay*-lax. *Diapensiaceae*. From
Gk. *gala* (milk), after the white flowers. Evergreen, perennial herbs.
   urceolata, ur-see-o-*la*-ta. Urn-
   shaped.

Galega, ga-*lee*-ga. *Leguminosae.* From Gk. *gala* (milk). Perennial herbs.
officinalis, o-fi-si-*na*-lis. Sold in shops.
orientalis, o-ree-en-*ta*-lis. Eastern.

Galeobdolon, ga-lee-*ob*-do-lon. *Labiatae.* From L. *galeo* (cover with a helmet) and *dolon* (fly's sting). Semi-evergreen perennial herbs.
luteum, loo-tee-um. Yellow (flowers).

Galeobdolon luteum

Galium, *ga*-lee-um. *Rubiaceae.* From Gk. *gala* (milk). *G. verum* (lady's bed-straw). Perennial herb.
longifolium, long-i-*fo*-lee-um. Long-leaved.
odoratum, o-do-*ra*-tum. Scented. Sweet Woodruff.

Galtonia, gawl-*ton*-ee-a. *Liliaceae.* After Sir Frances Galton. Bulbous herbs.
candicans, kan-di-kanz. White (flowers). Summer Hyacinth.
princeps, prin-seps. Distinguished.
viridiflora, vi-ri-di-*flo*-ra. With green flowers.

Gardenia, gar-*den*-ee-a. *Rubiaceae.* After Dr Alexander Garden. Tender, evergreen trees and shrubs.
capensis, ka-*pen*-sis. Of the Cape of Good Hope.
florida, flo-ri-da. Flowering.
jasminoides, jas-min-*oi*-deez. *Jasminum*-like (flowers).

Garrya, *ga*-ree-a. *Garryaceae.* After Nicholas Garry. Evergreen trees and shrubs.
elliptica, e-*lip*-ti-ka. Elliptic (leaves).

Gasteria, gas-*te*-ree-a. *Liliaceae.* From Gk. *gaster* (belly), after the swollen base of the corolla lube. Tender perennial succulents.
brevifolia, brev-i-*fo*-lee-a. Short-leaved.
caespitosa, see-spi-*to*-sa. Tufted.
liliputana, lil-ee-put-ee-*a*-na. Very small.
maculata, mak-ew-*la*-ta. Spotted (leaves).
marmorata, mar-mo-*ra*-ta. Marbled (leaves).
verrucosa, ve-roo-*ko*-sa. Warty (leaves). Wart Gasteria.

Gaultheria, gawl-*the*-ree-a. *Ericaceae.* After Dr Gaulthier. Evergreen shrubs and sub-shrubs.
forrestii, fo-*rest*-ee-ee. After George Forrest.
procumbens, pro-*kum*-benz. Prostrate. Creeping Wintergreen.
shallon, sha-lon. The native name. Sabal, Shallon.

Gazania, ga-*za*-nee-a. *Compositae.* After Theodore of Gaza. Semi-hardy perennial herbs.
pinnata, pin-*a*-ta. Pinnate.
rigens, ri-jens. Rigid. Treasure Flower.

leucolaena, loo-ko-*lee*-na. White-cloaked, white leaves.
uniflora, ew-ni-*flo*-ra. With one flower.

Gelsemium, gel-*sem*-ee-um. *Loganiaceae.* From the Italian *gelsomino* (jasmine). Semi-hardy, evergreen climbers.
sempervirens, sem-per-*vi*-rens. Evergreen. Yellow Jessamine.

Genista, je-*nis*-ta. *Leguminosae.* L. name. Deciduous shrubs. Broom.
aetnensis, eet-*nen*-sis. Of Mt Etna, Sicily. Mt Etna Broom.
albida, al-bi-da. White.
cinerea, sin-*e*-ree-a. Grey.
delphinensis, del-fin-*en*-sis. Of Dauphine, France.
fragrans, fra-granz. Fragrant.
lydia, li-dee-a. Of Lydia (Turkey).
pilosa, pi-*lo*-sa. With long, soft hairs.
tenera, te-ne-ra. Tender, delicate.
tinctoria, tink-*to*-ree-a. Used in dyeing. Dyer's Greenweed.

Gentiana, jen-tee-*a*-na. *Gentianaceae.* After Gentius, King of Illyria. Annual, biennial and perennial herbs. Gentian.
acaulis, a-*kaw*-lis. Without a stem. Trumpet Gentian.
andrewsii, an-*drooz*-ee-ee. After H. C. Andrews. Bottle Gentian.
angustifolia, ang-gus-ti-*fo*-lee-a. Narrow-leaved.
asclepiadea, a-sklee-pee-*a*-dee-a. *Asclepias*-like. Willow Gentian.
clusii, *klooz*-ee-ee. After Charles de l'Ecluse.
farreri, *fa*-ra-ree. After Farrer.
gracilipes, gra-*sil*-i-pees. Slender-stalked (flowers).
hexaphylla, hex-a-*fil*-a. Six-leaved.
lutea, *loo*-tee-a. Yellow (flowers). Yellow Gentian.
ornata, or-*na*-ta. Decorative.
pneumonanthe, new-mon-*anth*-ee. Lung flower. Marsh Gentian.
punctata, punk-*ta*-ta. Spotted (flowers).
pyrenaica, pi-ren-*ee*-i-ka. Of the Pyrenees.
saxosa, sax-*o*-sa. Growing in rocky places.

Genista pilosa

Gentiana pneumonanthe

sino-ornata, sie-no-or-*na*-ta. The Chinese *G. ornata.*
verna, ver-na. Of spring. Spring Gentian.

Geranium, jer-*ay*-nee-um. *Geraniaceae.* From Gk. *geranos* (crane), after the beak-like fruits. Perennial herbs. Cranesbill.
  argenteum, ar-*jen*-tee-um. Silvery.
  armenum, ar-*meen*-um. Of Armenia.
  cinereum, si-*ne*-ree-um. Grey (leaves).
  dalmaticum, dal-*mat*-i-kum. Of Dalmatia.
  farreri, fa-ra-ree. After Farrer.
  himalayense, hi-ma-lay-*en*-see. Of the Himalaya.
  lucidum, loo-si-dum. Glossy.
  macrorrhizum, mak-ro-*ree*-zum. With a large root.
  nervosum, ner-*vo*-sum. Veined.
  nodosum, no-*do*-sum. With conspicuous nodes.
  phaeum, fee-um. Dusky. Mourning Widow, Dusky Cranesbill.
  pratense, pra-*ten*-see. Of meadows.
  procurrens, pro-*ku*-renz. Spreading.

psilostemon, see-*lo*-ste-mon. With smooth stamens.
renardii, re-*nar*-dee-ee. After Charles Claude Renard.
sanguineum, sang-*gwin*-ee-um. Bloody. Bloody Cranesbill.
sylvaticum, sil-*va*-ti-kum. Of woods. Wood Cranesbill.
wallichianum, wo-lik-ee-*a*-num. After Wallich.

Gerbera, *ger*-ba-ra. *Compositae.* After Traugott Gerber. Semi-hardy perennial herbs.
  jamesonii, jaym-*son*-ee-ee. After Jameson. Barberton Daisy.

Geum, *jee*-um. *Rosaceae.* L. name. Perennial herbs.
  borisii, bo-*ris*-ee-ee. After Boris.
  bulgaricum, bul-*ga*-ri-kum. Of Bulgaria.
  chiloense, chi-lo-*en*-se. Of Chiloe.
  montanum, mon-*ta*-num. Of mountains.
  reptans, rep-tanz. Creeping.
  rivale, ri-*va*-lee. Growing by streams. Water Avens.

Geranium sylvaticum

Geum rivale

Gilia, *gi*-lee-a. *Polemoniaceae.* After Filippo Luigi Gilii. Annual herbs.
achilleifolia, a-ki-lee-i-*fo*-lee-a. *Achillea*-leaved.
capitata, ka-pi-*ta*-ta. In a dense head (flowers). Blue Thimble Flower.
tricolor, tri-ko-lor. Three-coloured. Bird's Eyes.

Gillenia, gi-*len*-ee-a. *Rosaceae.* After Arnold Gillenius. Perennial herb.
trifoliata, tri-fo-lee-*a*-ta. With three leaves.

Ginkgo, *gink*-go. *Ginkgoaceae.* From Japanese *ginkyo* (Silver Apricot). Deciduous tree.
biloba, bi-*lo*-ba. Two-lobed (leaves). Maidenhair Tree.

Gladiolus, gla-*dee*-o-lus. *Iridaceae* From L. for small sword, after the leaves. Cormous perennial herbs.
blandus, bland-us. mild.
byzantinus, bi-zan-*teen*-us. Of Byzantium (Istanbul).
cardinalis, kar-di-*na*-lis. Scarlet (flowers).
carneus, kar-nee-us. Flesh-coloured (flowers).
colvillii, kol-*vil*-ee-ee. After James Colvill.
gandavensis, gan-da-*ven*-sis. Of Ghent.
hortulanus, hort-ew-*la*-nus. Of gardens.
illyricus, i-*li*-ri-kus. Of Illyria.
imbricatus, im-bri-*ka*-tus. Overlapping.
liliaceus, lil-ee-*a*-see-us. *Lilium*-like.
natalensis, na-ta-*len*-sis. Of Natal.
primulinus, prim-ew-*leen*-us. Primrose yellow.
recurvus, re-*kur*-vus. Curved back.
tristis, tris-tis. Sad.
undulatus, un-dew-*la*-tus. Wavy-margined.

Glaucidium, glaw-*sid*-ee-um. *Paeoniaceae.* From *Glaucium.* Perennial herb.
palmatum, parl-*ma*-tum. Hand-like (leaves).

Glaucium, *glaw*-see-um. *Papaveraceae.* From Gk. *glaukos* (grey-green), after the colour of the leaves. Annual, biennial or perennial herbs.
corniculatum, kor-nik-ew-*la*-tum. Horned. Red Horned Poppy.
flavum, fla-vum. Yellow. Yellow Horned Poppy.

Glechoma, gle-*ko*-ma. *Labiatae.* From Gk. *glechon* (mint), after the scent of the leaves. Evergreen perennial herbs.
hederacea, he-de-*ra*-see-a. *Hedera*-like. Ground Ivy.

Glechoma hederacea

Gleditsia, gled-*it*-she-a. *Leguminosae.* After Gottlieb Gleditsch. Deciduous trees.
caspica, kas-pi-ka. Of the region of the Caspian Sea. Caspian Locust.
japonica, ja-*pon*-i-ka. Of Japan.

triacanthos, tri-a-*kanth*-os. Three-spined. Honey Locust.

Globularia, glob-ew-*la*-ree-a. *Globulariaceae.* From L. *globulus* (small ball), after the flower heads. Evergreen shrubs and sub-shrubs. Globe Daisy.
  cordifolia, kor-di-*fo*-lee-a. With heart-shaped leaves.
  meridionalis, me-ree-dee-o-*na*-lis. Flowering at mid-day.
  nudicaulis, new-di-*kaw*-lis. Bare-stemmed.

Gloriosa, glo-ree-*o*-sa. *Liliaceae.* From L. *gloriosus* (glorious). Tender, deciduous climbers. Glory Lily.
  rothschildiana, roths-child-ee-*a*-na. After Lionel Walter, 2nd Baron Rothschild.
  superba, soo-*per*-ba. Superb.

Gloxinia, glox-*in*-ee-a. *Gesneriaceae.* After Benjamin Gloxin. Tender perennial herb.
  perennis, pe-*ren*-is. Perennial.
  speciosa, spes-ee-*o*-sa. Showy.

Glyceria, gli-*se*-ree-a. *Gramineae.* From Gk. *glykis* (sweet). Some seeds are edible. Perennial Grass.
  aquatica, a-*kwa*-ti-ka. Growing in water.
  maxima, max-i-ma. Largest. Reed Grass.

Gomphrena, gom-*free*-na. *Amaranthaceae.* L. name. Annual, biennial and perennial herbs.
  globosa, glo-*bo*-sa. Spherical (flower heads). Globe Amaranth.

Gordonia, gor-*don*-ee-a. *Theaceae.* After James Gordon. Evergreen trees and shrubs.

axillaris, ax-il-*la*-ris. In the leaf axil.
pubescens, pew-*bes*-senz. Hairy.

Gossypium, go-*sip*-ee-um. *Malvaceae.* From L. *gossypion.* Tender shrubs, sub-shrubs and herbs.
  arboreum, ar-*bor*-ee-um. Tree-like. Tree Cotton.
  barbadense, bar-bad-*en*-se. Of Barbados.
  herbaceum, her-*ba*-see-um. Herbaceous. Levant Cotton.

Graptopetalum, grap-to-*pe*-ta-lum. *Crassulaceae.* From Gk. *graptos* (written upon) and *petalon* (petal), the markings on the petals. Tender perennial succulents.
  amethystinum, a-me-*this*-ti-num. Violet (leaves).
  bellum, be-lum. Pretty.
  pachyphyllum, pa-kee-*fil*-lum. Thick-leaved.
  paraguayense, pa-ra-gwie-*en*-see. Of Paraguay. Ghost Plant, Mother of Pearl Plant.

Grevillea, gre-*vil*-ee-a. *Proteaceae.* After Charles F. Greville. Tender, semi-hardy and evergreen trees and shrubs.
  robusta, ro-*bus*-ta. Robust. Silky Oak.
  rosmarinifolia, ros-ma-reen-i-*fo*-lee-a. *Rosmarinus*-leaved.
  sulphurea, sul-*fu*-ree-a. Sulphur-yellow (flowers).

Grindelia, grin-*del*-ee-a. *Compositae.* After David H. Grindel. Evergreen, perennial herbs and sub-shrubs.
  speciosa, spes-ee-*o*-sa. Showy.

Griselinia, gri-se-*leen*-ee-a. *Cornaceae.* After Francesco Griselini. Semi-hardy, evergreen trees or shrubs.

littoralis, li-to-*ra*-lis. Of the sea shore.
lucida, loo-si-da. Glossy.

Gunnera, gun-*e*-ra. *Gunneraceae.*
After Ernst Gunnerus. Perennial herbs.
chilensis, chi-*len*-sis. Of Chile.
magellanica, ma-jel-*an*-ik-a. From
the region of the Magellan Straits.
manicata, man-i-*ka*-ta. With long
sleeves.

Guzmania, guz-*man*-ee-a.
Bromeliaceae. After Anastasio
Guzman. Tender, evergreen perennials.
lingulata, ling-gew-*la*-ta. Tongue-
like (bracts).
monostachia, mon-o-*stak*-ee-a. With
one spike.
musaica, mu-*sa*-i-ka. Banana-like.
(leaves).
sanguinea, sang-*gwin*-ee-a. Red.
tricolor, tri-ko-lor. Three-coloured

Gymnocalycium, jim-no-ka-*li*-see-
um. *Cactaceae.* From Gk. *gymnos*
(naked) and *calyx* (bud).
gibbosum, ji-*bo*-sum. Swollen on
one side.
mihanovichii, mi-han-no-*vich*-ee-ee.
After Mihanovich.
multiflorum, mul-ti-*flo*-rum. Many-
flowered.

platense, pla-*ten*-see. From near the
River Plate, Argentina.
quehlianum, kwel-ee-*a*-num. After
Leopold Quehl.

Gymnocladus, gim-*no*-kla-dus.
Leguminosae. From Gk. *gymnos*
(naked) and *klados* (branch).
Deciduous tree.
dioica, dee-o-*ee*-ka. Dioecious.
Kentucky Coffee Tree.

Gynura, gin-*ew*-ra *Compositae.* From
Gk. *gyne* (female) and *oura* (tail), after
the stigma. Tender, evergreen perennial herbs and climbers.
aurantiaca, aw-ran-tee-*a*-ka.
Orange (flowers). Velvet Plant,
Purple Passion Vine.

Gypsophila, jip-*sof*-i-la.
Caryophyllaceae. From Gk. *gypsos*
(gypsum) and *philos* (loving). Annual
and perennial herbs.
cerastioides, se-ras-tee-*oi*-deez.
Cerastium-like.
elegans, e-le-ganz. Elegant.
paniculata, pa-nik-ew-*la*-ta. With
flowers in panicles.
petraea, pe-*tree*-a. Growing on
rocks.
repens, ree-penz. Creeping.

H

Haageocereus, harg-ee-o-*see*-ree-us. *Cactaceae.* After J. N. Haage and *Cereus.*
  decumbens, dee-*kum*-benz. Prostrate.
  versicolor, ver-*si*-ko-lor. Variably coloured (spines).

Haberlea, ha-*ber*-lee-a. *Gesneriaceae.* After Carl Constantin Haberle. Evergreen, perennial herbs.
  ferdinandii-coburgii, fer-di-*nan*-dee-ee-ko-*burg*-ee-ee. After King Ferdinand of Bulgaria.
  rhodopensis, ro-do-*pen*-sis. Of the Rhodope Mountains, Bulgaria.

Habranthus, ha-*branth*-us. *Amaryllidaceae.* From Gk. *habros* (graceful) and *anthos* (flower). Hardy to semi-hardy bulbous herbs.
  andersonii, an-der-*son*-ee-ee. After Anderson.
  robustus, ro-*bus*-tus. Robust.

Haemanthus, heem-*anth*-us. *Amaryllidaceae.* From Gk. *haima* (blood) and *anthos* (flower), after the colour of the flowers. Tender bulbous herbs. Blood Lily, Red Cape Tulip.
  coccineus, kok-*kin*-ee-us. Scarlet. Ox-tongue Lily.
  humilis, *hum*-i-lis. Low-growing.
  katharinae, kath-a-*rin*-ee. After Mrs Katherine Saunders. Blood Flower.
  magnificus, mag-*ni*-fi-kus. Magnificent. Royal Paint Brush.
  multiflorus, mul-ti-*flo*-rus. Many-flowered. Salmon Blood Lily.
  puniceus, pew-*ni*-see-us. Reddish-purple.
  sanguineus, sang-*win*-ee-us. Blood-red.

Hakea, *hak*-ee-a. *Proteaceae.* After Baron Christian Ludwig von Hake. Hardy and tender, evergreen trees and shrubs.
  laurina, law-*reen*-a. *Laurus*-like. Pincushion Flower, Sea Urchin.
  lissosperma, lis-o-*sperm*-a. With smooth seeds.
  microcarpa, mie-kro-*kar*-pa. With small fruits.
  suaveolens, swa-vee-*o*-lenz. Sweet-scented.

Halesia, *haylz*-ee-a. *Styracaceae.* After Dr Stephen Hales. Deciduous trees and shrubs. Silver Bell, Snowdrop Tree.
  carolina, ka-ro-*leen*-a. Of Carolina.
  diptera, *dip*-te-ra. Two-winged (fruit).
  magniflora, mag-ni-*flo*-ra. Large-flowered.
  monticola, mon-*ti*-ko-la. Mountain-loving. Mountain Snowdrop Tree.

X Halimiocistus, ha-lim-ee-o-*sis*-tus. *Cistaceae.* Hybrids (*Cistus* X *Halimium*). Evergreen shrubs.
  sahucii, sa-*hook*-ee-ee. After M. Sahuc.
  wintonensis, win-ton-*en*-sis. Of Winchester.

Halimium, ha-*lim*-ee-um. *Cistaceae.* From *Atriplex halimus.* Evergreen shrubs.
  lasianthum, la-see-*anth*-um. With woolly flowers.
  formosum, for-*mo*-sum. Beautiful.
  ocymoides, o-kim-*oi*-deez. Like *Ocimum.*
  umbellatum, um-bel-*a*-tum. With flowers in umbels.

Hamamelis, ham-a-*mee*-lis. *Hamamelidaceae.* From Gk. hama (together) and mela (fruit). Deciduous shrubs. Witch Hazel.
  japonica, ja-*pon*-i-ka. Of Japan. Japanese Witch Hazel.
  mollis, mol-lis. Softly hairy (shoots and leaves). Chinese Witch Hazel.
  vernalis, ver-*na*-lis. Of spring.

Hardenbergia, hard-en-*berg*-ee-a. *Leguminosae.* After Countess von Hardenberg. Tender evergreen climber.
  comptoniana, komp-ton-ee-*a*-na. After Compton, the family name of Lady Northampton.
  violacea, vie-o-*la*-see-a. Violet.

Hatiora, ha-tee-*o*-ra. *Cactaceae.* After Thomas Hariot. Cacti.
  bambusoides, bam-bew-*soi*-deez. Bamboo-like.
  salicornioides, sa-li-korn-ee-*oi*-deez. *Salicornia*-like. Bottle Cactus.

Haworthia, hay-*werth*-ee-a. *Liliaceae.* After Adrian Hardy Haworth. Tender, perennial succulents.
  attenuata, a-ten-ew-*a*-ta. Drawn out (leaves).
  cuspidata, kus-pi-*da*-ta. With a sharp, short point (leaves).
  margaritifera, mar-ga-ri-*ti*-fe-ra. Pearl-bearing (leaves). Pearl Plant.
  reinwardtii, rien-*vart*-ee-ee. After Reinwardt.
  tesselata, te-se-*la*-ta. Chequered. Wart Plant.
  truncata, trun-*ka*-ta. Abruptly cut off (leaves).

Hebe, *hee*-bee. *Scrophulariaceae.* After Hebe, goddess of youth. Evergreen shrubs.
  albicans, al-bi-kanz. Whitish (leaves).
  andersonii, an-der-*son*-ee-ee. After

Isaac Anderson-Henry.
  brachysiphon, bra-kee-*see*-fon. With a short tube (corolla).
  chathamica, cha-*tam*-i-ka. Of the Chatham Islands.
  cupressoides, kew-pres-*oi*-deez. *Cupressus*-like.
  franciscana, fran-sis-*ka*-na. Of San Francisco.
  glaucophylla, glaw-ko-*fil-a*. With glaucous leaves.
  hulkeana, hulk-ee-*a*-na. After T. H. Hulke. New Zealand Lilac.
  macrantha, ma-*kranth*-a. Large-flowered.
  ochracea, ok-*ra*-see-a. Ochre-coloured (foliage).
  pinguifolia, pin-gwi-*fo*-lee-a. With fat leaves.
  rakaiensis, ra-kie-*en*-sis. Of the Rakai Valley, Canterbury.
  recurva, re-*kur*-va. Curved back (leaves).
  salicifolia, sa-li-si-*fo*-lee-a. *Salix*-leaved.

Hedera, *he*-de-ra. *Araliaceae.* L. name. Evergreen, perennial herbs and

Hedera helix

climbers. Ivy.
algeriensis, al-je-ree-*en*-sis. Of
Algeria.
azorica, a-*zo*-ri-ka. Of the Azores.
colchica, kol-chi-ka. Of Colchis,
Black Sea.
dentata, den-*ta*-ta. Toothed (leaves).
helix, he-lix. Entwining. Common
Ivy.
poetica, po-*et*-i-ka. Of poets. Poet's
Ivy.
hibernica, hi-*bern*-i-ka. Of Ireland.
Irish Ivy.

Hedychium, he-*di*-kee-um.
Zingiberaceae. From Gk. *hedys*
(sweet) and *chion* (snow). Semi-hardy
to tender, perennial herbs. Ginger Lily.
coronarium, ko-ro-*na*-ree-um. Used
in garlands. Butterfly Lily, Garland
Flower.
densiflorum, dens-i-*flo*-rum. Densely
flowered. Temperate.
flavescens, fla-*ves*-enz. Yellow.
gardnerianum, gard-na-ree-*a*-num.
After Edward Gardner.

Hedyotis, he-*dee*-o-tis. *Rubiaceae.*
From Gk. *hedys* (sweet) and *otos* (ear).
Perennial herbs.
caerulea, see-*ru*-lee-a. Dark blue.
Bluets.
michauxii, mee-*sho*-ee-ee. After
Andre Michaux.
purpurea, pur-*pur*-ree-a. Purple.

Hedysarum, he-*dis*-a-rum.
Leguminosae. From Gk. *hedys* (sweet).
Perennial and biennial herbs and
deciduous sub-shrubs.
apiculatum, a-pik-ew-*la*-tum. With
an abrupt, short point.
coronarium, ko-ro-*na*-ree-um. Used
in garlands. French Honeysuckle.

Helenium, he-*le*-nee-um. *Compositae.*
After Helen of Troy. Perennial herbs.

Sneezeweed.
autumnale, aw-tum-*na*-lee. Of
autumn (flowering).

Helianthemum, hee-lee-*anth*-e-mum.
Cistaceae. From Gk. *helios* (sun) and
anthemon (flower). Evergreen shrubs.
Rock Rose.
alpestre, al-*pes*-tree. Of the lower
mountains.
appeninum, a-pe-*neen*-um. Of the
Apennines.
lunulatum, loon-ew-*la*-tum.
Crescent-shaped.
oelandicum, ur-land-*i*-kum. Of
Oland, Sweden.

Helianthus, hee-lee-*anth*-us.
Compositae. From Gk. *helios* (sun)
and *anthos* (flower). Annual and
perennial herbs.
annuus, an-ew-us. Annual.
Sunflower.
decapetalus, dek-a-*pe*-ta-lus. With
ten petals.
multiflorus, mul-ti-*flo*-rus. Many-
flowered.
tuberosus, tew-be-*ro*-sus. Tuberous
(rhizome). Jerusalem Artichoke.

Helichrysum, hee-li-*kris*-um.
Compositae. From Gk. *helios* (sun)
and *chryson* (golden). Annual and
perennial herbs and evergreen shrubs
and sub-shrubs.
angustifolium, an-gus-ti-*fo*-lee-um.
Narrow-leaved.
bellidioides, bel-i-dee-*oi*-deez.
Bellis-like .
bracteatum, brak-tee-*a*-tum. With
conspicuous bracts. Everlasting.
italicum, i-*ta*-li-kum. Of Italy.
milfordiae, mil-*ford*-ee-ee. After Mrs
Helen Milford.
orientale, o-ree-en-*ta*-lee. Eastern.
petiolare, pe-tee-o-*la*-ree. With con-

spicuous petioles.
plicatum, pli-*ka*-tum. Pleated (leaves).
serotinum, se-*ro*-ti-num. Late flowering. Curry Plant.
sibthorpii, sib-*thorp*-ee-ee. After John Sibthorp.
splendidum, *splen*-di-dum. Splendid.

Helictotrichon, he-lik-to-*tri*-kon.
Gramineae. From Gk. *helix* (spiral) and *trichos* (hair). Perennial grass.
sempervirens, sem-per-*vi*-rens. Evergreen.

Heliocereus, hee-lee-o-*see*-ree-us.
Cactaceae. From Gk. *helios* (sun) and *Cereus.*
cinnabarinus, sin-a-ba-*reen*-us. Cinnabar red (flowers).
speciosus, spes-ee-*o*-sus. Showy.

Heliophila, hee-lee-*o*-fi-la. *Cruciferae.*
From Gk. *helios* (sun) and *philos* (loving). Annual and perennial herbs and sub-shrubs. Cape Stock.
africana, af-ri-*ka*-na. African.
leptophylla, lep-to-*fil*-a. Narrow-leaved.
linearifolia, lin-ee-a-ri-*fo*-lee-a. With narrow leaves.

Heliopsis, hee-lee-*op*-sis. *Compositae.*
From Gk. *helios* (sun) and -*opsis* (resemblance). Perennial herbs. Ox Eye.
helianthoides, hee-lee-anth-*oi*-deez. *Helianthus*-like.

Heliotropium, hee-lee-o-*tro*-pee-um.
Boraginaceae. From Gk. *helios* (sun) and *trope* (turn). Annual herbs and evergreen shrubs.
arborescens, ar-bo-*res*-enz. Becoming tree-like. Heliotrope, Cherry Pie.
indicum, *in*-di-kum. Of India.

Helipterum, hee-*lip*-te-rum.
Compositae. From Gk. *helios* (sun) and *pteron* (wing), after the plumed pappus. Annual and perennial herbs.
canescens, ka-*nes*-enz. Greyish-white hairs.
manglesii, man-*galz*-ee-ee. After Mangles.
roseum, ro-see-um. Rose-coloured (flower bracts).

Helleborus viridis

Helleborus, he-*le*-bo-rus.
Ranunculaceae. From Gk. *helleboros.*
Perennial herbs. Hellebore.
argutifolius, ar-gew-ti-*fo*-lee-us. With sharply-toothed leaves.
atrorubens, at-ro-*ru*-benz. Deep red.
foetidus, fee-ti-dus. Stinking. Stinking Hellebore.
lividus, li-vi-dus. Lead-coloured.
niger, ni-ger. Black (roots). Christmas Rose.
orientalis, o-ree-en-*ta*-lis. Eastern. Lenten Rose.
purpurascens, pur-pur-*ras*-enz. Purplish.
sternii, stern-ee-ee. After Sir Frederick Stern.
viridis, vi-ri-dis. Green (sepals). Green Hellebore.

Hemerocallis, hee-me-ro-*ka*-lis.
Liliaceae. From Gk. *hemera* (day) and
kallos (beauty). The flowers last one
day only. Perennial herbs. Day Lily.
   citrina, si-*tree*-na. Lemon-yellow.
   dumortieri, dew-mor-tee-*e*-ree. After
   B. C. Dumortier.
   fulva, ful-va. Tawny.
   middendorfii, mid-an-*dorf*-ee-ee.
   After Alexander Theodor von
   Middendorf.
   minor, mi-nor. Smaller.
   multiflora, mul-ti-*flo*-ra. Many-flowered.

Hemigraphis, he-mi-*graf*-is.
Acanthaceae. From Gk. *hemi* (half)
and *graphis* (brush), after the hairy filaments of the stamens. Annual or
perennial herbs and sub-shrubs.
   alternata, al-ter-*na*-ta. Alternate.
   Red Ivy.

Hepatica, he-*pa*-ti-ka. *Ranunculaceae.*
From Gk. *hepar* (liver), after the leaf
shape. Perennial herbs.
   americana, a-me-ri-*ka*-na. Of
   America.
   nobilis, no-bi-lis. Notable.
   transsilvanica, trans-sil-*va*-ni-ka. Of
   Transylvania, Romania.

Heracleum, he-ra-*klee*-um.
Umbelliferae. After Hercules. Biennial
or perennial herbs.
   giganteum, ji-*gan*-te-um. Gigantic.
   mantegazzianum, man-tee-gatz-ee-*a*-
   num. After Paolo Mantegazzi. Giant
   Hogweed, Cartwheel Flower.

Hermodactylus, her-mo-*dak*-ti-lus.
Iridaceae. From Gk. Hermes and *dak-*
tylos (finger), after the finger-like
tubers. Tuberous, perennial herbs.
   tuberosus, tew-be-*ro*-sus. Tuberous.
   Snake's-head Iris.

Herniaria, her-nee-*a*-ree-a.
Caryophyllaceae. From L. *hernia*
(rupture), after its alleged healing
properties. Annual or perennial herbs.
   glabra, gla-bra. Smooth.
   Rupturewort.

Herniaria glabra

Hesperis, *hes*-pe-ris. *Cruciferae.* From
Gk. *hespera* (evening), after the fragrant evening flowers. Annual or perennial herbs.
   matronalis, ma-tro-*na*-lis. Of
   matrons. Sweet Rocket, Dame's
   Violet.

Hesperis matronalis

Heterocentron, het-er-o-*sen*-tron. *Melastomataceae.* From Gk. *heteros* (different) and *kentron* (spur). Tender, perennial herbs and sub-shrubs.
elegans, e-le-ganz. Elegant. Spanish Shawl.

Heuchera, *hoy*-ka-ra. *Saxifragaceae.* After Johann Heinrich von Heucher. Evergreen perennial herbs. Alum Root.
sanguinea, sang-*gwin*-ee-a. Blood-red (flowers).
americana, a-mer-ik-*a*-na. Of America. Rock Geranium.

Hibbertia, hi-*bert*-ee-a. *Dilleniaceae.* Tender, evergreen shrubs and climbers. Button Flower.
dentata, den-*ta*-ta. Toothed (leaves).
scandens, skan-denz. Climbing. Gold Guinea Plant, Snake Vine.

Hibiscus, hi-*bis*-kus. *Malvaceae.* Gk. for mallow. Tender, annual and perennial herbs; hardy and tender trees and shrubs.
moscheutos, mos-*kew*-tas. Musk-scented. Swamp Rose Mallow.
mutabilis, mew-*ta*-bi-lis. Changeable (flower colour). Cotton Rose.
rosa-sinensis, ro-sa-si-*nen*-sis. China Rose.
schizopetalus, ski-zo-*pe*-ta-lus. With split petals. Japanese Hibiscus.
syriacus, si-ree-*a*-kus. Of Syria.
trionum, tri-o-num. Three-coloured. Flower of an Hour.

Hieracium, hee-e-*ra*-see-um. *Compositae.* From Gk. *hierax* (hawk). Perennial herbs.
aurantiacum, aw-ran-tee-*a*-kum. Orange (flowers). Devil's Paintbrush.
maculatum, mak-ew-*la*-tum. Spotted (leaves).
villosum, vi-*lo*-sum. Softly hairy.

Hieracium aurantiacum

Hippeastrum, hip-ee-*as*-trum. *Amaryllidaceae.* From Gk. *hippos* (horse). Tender bulbous herbs. Amaryllis.
argentinum, ar-jen-*teen*-um. Of Argentina.
aulicum, aw-li-kum. Of the court. Lily of the Palace.
puniceum, pew-*ni*-see-um. Reddish-purple (flowers).
reticulatum, ree-tik-ew-*la*-tum. Net-veined (perianth lobes).
striatum, stri-*a*-tum. Striped (flowers).
vittatum, vi-*ta*-tum. Banded (flowers).

Hippocrepis, hi-po-*kre*-pis. *Leguminosae.* From Gk. *hippos* (horse) and *krepis* (shoe), after the horseshoe-shaped pod segments. Annual and perrenial herbs.
comosa, ko-*mo*-sa. Tufted. Horseshoe Vetch.

Hippophae, hi-*po*-fa-ee. *Elaeagnaceae.* From the Gk. name. Deciduous trees and shrubs.
rhamnoides, ram-*noi*-deez. *Rhamnus*-like. Sea Buckthorn.

Hoheria, ho-*he*-ree-a. *Malvaceae.*
From the Maori *houhere.* Semi-hardy,
evergreen and deciduous trees and
shrubs.
angustifolia, an-gust-i-*fo*-lee-a.
Narrow-leaved.
glabrata, glab-*ra*-ta. Rather smooth.
lyallii, lie-*al*-ee-ee. After David
Lyall.
populnea, po-*pul*-nee-a. *Populus*-
like (leaves). Lacebark.
sexstylosa, sex-sti-*lo*-sa. With six
styles. Ribbonwood.

Holboellia, hol-*burl*-ee-a.
Lardizabalaceae. After F. L. Holboell.
Evergreen climber.
coriacea, ko-ree-*a*-see-a. Leathery
(leaflets).

Holcus, *hol*-kus. *Gramineae.* From the
Gk. *holkus.* Perennial grass.
mollis, mol-lis. Softly hairy.

Holcus mollis

Holmskioldia, holm-*shol*-dee-a.
Verbenaceae. After Theodor
Holmskiold. Tender, evergreen shrubs
and climbers.

sanguinea, sang-*gwin*-ee-a. Blood-
red (flowers). Chinese Hat Plant.

Holodiscus, ho-lo-*dis*-kus. *Rosaceae.*
From Gk. *holos* (entire) and *diskos*
(disc). Deciduous shrub.
discolor, dis-ko-lor. Two-coloured.
Ocean Spray.

Hordeum, *hor*-dee-um. *Gramineae.*
The L. for barley. Perennial grass.
jubatum, joo-*ba*-tum. Maned.
Squirrel-tail Grass.

Horminum, hor-*meen*-um. *Labiatae.*
The Gk. for sage. Perennial herb.
pyrenaicum, pi-ren-*ee*-i-kum. Of the
Pyrenees. Dragon's Mouth.

Hosta, *host*-a. *Liliaceae.* After Nicolas
Tomas Host. Perennial herbs. Funkia,
Plantain Lily.
crispula, krisp-ew-la. Wavy-margined.
decorata, de-ko-*ra*-ta. Decorative.
elata, e-*la*-ta. Tall.
fortunei, for-*tewn*-ee-ee. After
Robert Fortune.
rectifolia, rek-ti-*fo*-lee-a. With erect
leaves.
sieboldiana, see-bold-ee-*a*-na. After
Siebold.
tardiflora, tar-di-*flo*-ra. Late flowering.
undulata, un-dew-*la*-ta. Wavy-margined.
ventricosa, ven-tri-*ko*-sa. Swollen on
one side.

Hottonia, ho-*ton*-ee-a. *Primulaceae.*
After Peter Hotton. Deciduous, perennial, aquatic herbs.
inflata, in-*fla*-ta. Swollen. American
Featherfoil.
palustris, pa-*lus*-tris. Growing in
marshes. Water Violet.

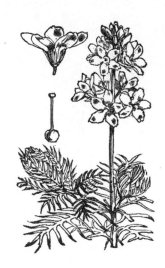

Hottonia palustris

Houttuynia, hoo-*tie*-nee-a.
Saururaceae. After Martin Houttuyn.
Perennial herbs.
   cordata, kor-*da*-ta. Heart-shaped
   (leaves).

Howea, *how*-ee-a. *Palmae.* After Lord
Howe. Tender, evergreen palms.
Sentry Palm, Paradise Palm.
   belmoreana, bel-mor-ree-*a*-na. After
   De Belmore. Curly Sentry Palm.
   forsteriana, for-sta-ree-*a*-na. After
   William Forster. Sentry Palm.

Hoya, *hoy*-a. *Asclepiadaceae.* After
Thomas Hoy. Tender, evergreen
climbers and shrubs.
   australis, aw-*stra*-lis. Southern.
   bella, be-la. Pretty.
   carnosa, kar-no-sa. Fleshy. Honey
   Plant. Wax Plant.
   imperialis, im-peer-ee-*a*-lis. Showy.

Huernia, hoo-*ern*-ee-a,
Asclepiadaceae. After Justin Heurnius.
Tender, perennial succulents.

keniensis, ken-ee-*en*-sis. Of Kenya.
Kenyan Dragon Flower.
macrocarpa, mak-ro-*kar*-pa. Large-
fruited.
zebrina, ze-*breen*-a. Striped. Little
Owl.

Humulus, *hum*-ew-lus. *Cannabaceae.*
Perennial climbers.
   japonicus, ja-*pon*-i-kus. Of Japan.
   Japanese Hop.
   lupulus, lup-ew-lus. A small wolf.
   Common Hop.

Hunnemannia, hun-ee-*man*-ee-a.
Papaveraceae. After John Hunneman.
Semi-hardy perennial or annual herb.
   fumariifolia, few-ma-ree-i-*fo*-lee-a.
   With *Fumaria*-like leaves. Mexican
   Tulip Poppy.

Hyacinthoides, hi-a-sinth-*oi*-deez.
Liliaceae. From *Hyacinthus* and Gk. -
oides (resemblance). Bulbous perennial herbs.
   hispanica, his-*pa*-ni-ka. Of Spain.
   Spanish Bluebell.

Hyacinthoides non-scripta

non-scripta, non-*skrip*-ta.
Unmarked. Bluebell.

Hyacinthus, hi-a-*sinth*-us. *Liliaceae.*
After Hyakinthos, the beautiful
Spartan. Bulbous herb.
  amethystinus, a-me-*this*-ti-nus.
  Violet.
  azureus, a-*zew*-ree-us. Sky-blue.
  orientalis, o-ree-en-*ta*-lis. Eastern.
  Dutch Hyacinth.

Hydrangea, hi-*dran*-jee-a.
Hydrangeaceae. From Gk. *hydor*
(water) and *aggos* (jar), after the cup-
shaped fruits. Deciduous shrubs and
deciduous or evergreen climbers.
  arborescens, ar-bo-*res*-enz.
  Becoming tree-like.
  aspera, a-*spe*-ra. Rough (leaves).
  discolor, dis-ko-lor. Two coloured
  leaves.
  heteromalla, het-e-*ro*-mal-la.
  Variably hairy.
  involucrata, in-vo-loo-*kra*-ta. With
  an involucre.
  macrophylla, mak-ro-*fil-a*. Large-
  leaved.
  paniculata, pa-nik-ew-*la*-ta. With
  flowers in panicles.
  petiolaris, pe-tee-o-*la*-ris. With conspicuous petioles.
  quercifolia, kwer-ki-*fo*-lee-a.
  Quercus-leaved. Oak-leaved
  Hydrangea.
  sargentiana, sar-jen-tee-*a*-na. After
  Sargent.
  serrata, se-*ra*-ta. Saw-toothed
  (leaves).
  serratifolia, se-ra-ti-*fo*-lee-a. With
  saw-toothed leaves.

Hydrocharis, hi-*dro*-ka-ris.
Hydrocharitaceae. From Gk. *hydor*
(water) and *charis* (grace). Deciduous,
perennial, aquatic herbs.

Hydrocharis morsus-ranae

  morsus-ranae, mor-sus-*ra*-nee.
  Frog's bite. Frogbit.

Hygrophila, hi-*gro*-fi-la. *Acanthaceae.*
From Gk. *hygros* (moist) and *philos*
(loving), growing in wet places.
Deciduous or evergreen perennial
aquatic plants.
  difformis, di-*for*-mis. Of dissimilar
  shapes (leaves). Water Wisteria.
  polysperma, pol-i-*sperm*-a. With
  many seeds.

Hylocereus, hi-lo-*see*-ree-us.
Cactaceae. From Gk. *hyle* (wood) and
Cereus. Climbing cacti.
  trigonus, tri-*go*-nus. Three angled
  (stem).
  undatus, un-*da*-tus. Wavy (stem
  wings).

Hymenanthera, hi-men-an-*the*-ra.
Violaceae. From Gk. *hymen* (membrane) and *anthera* (anther). Semi-
hardy, evergreen shrubs.
  alpina, al-*pie*-na. Alpine.
  angustifolia, an-gust-i-*fo*-lee-a.
  Narrow-leaved.

crassifolia, kras-i-*fo*-lee-a. Thick-leaved.

Hymenocallis, hi-men-o-*kal*-is. *Amaryllidaceae*. From Gk. *hymen* (membrane) and *kallos* (beauty). Tender or semi-hardy bulbous herbs. Spider Lily.
  caribaea, ka-ri-*bee*-a. Of the Caribbean.
  festalis, fe-*sta*-lis. Festive.
  harrisiana, ha-ris-ee-*a*-na. After Mr T. Harris.
  littoralis, li-to-*ra*-lis. Of the shore.
  macrostephana, mak-ro-ste-*fa*-na. Large-crowned.
  narcissiflora, nar-sis-i-*flo*-ra. *Narcissus*-flowered. Peruvian Daffodil.

Hymenophyllum, hi-men-o-*fil*-lum. *Hymenophyllaceae*. From Gk. *hymen* (membrane) and *phyllon* (leaf). Terrestrial and epiphytic ferns. Filmy Ferns.
  tunbrigense, tun-brij-*en*-see. Of Tunbridge Wells. Tunbridge Wells Filmy Fern.
  wilsonii, wil-*son*-ee-ee. After Wilson.

Hymenophyllum tunbrigense

Hypericum, hi-per-*ee*-kum. *Hypericaceae*. Perennial herbs and deciduous or evergreen shrubs.
  androsaemum, an-dros-*ee*-mum. Blood-red sap.
  beanii, been-ee-ee. After Bean.
  calycinum, kal-i-*see*-num. Calyx-shaped.
  cerastioides, se-ras-tee-*oi*-deez. *Cerastium*-like.
  coris, ko-ris. *Coris*-like.
  empetrifolium, em-pet-ri-*fo*-lee-um. *Empetrum*-like.
  oliganthum, o-lig-*anth*-um. With few flowers.
  forrestii, fo-*rest*-ee-ee. After Forrest.
  olympicum, o-*lim*-pi-kum. Of Mt Olympus.
  reptans, *rep*-tanz. Creeping.
  setosum, see-*to*-sum. Bristly.

Hypericum androsaemum

Hypoestes, hi-po-*es*-teez. *Acanthaceae*. From Gk. *hypo* (below) and *estia* (house). Tender, evergreen, perennial herbs.
  phyllostachya, fil-lo-*stak*-ee-a. With leafy spikes. Polka-dot Plant, Baby's Tears, Pink Dot.

Hypoxis, hi-*pox*-is. *Hypoxidaceae.*
Perennial herbs.
angustifolia, an-gust-i-*fo*-lee-a.
Narrow-leaved.
capensis, ka-*pen*-sis. Of the Cape of
Good Hope. White Star Grass.
decumbens, dee-*kum*-benz. Prostrate.
elata, e-*la*-ta. Tall.
hirsuta, hir-*soo*-ta. Hairy (leaves).
latifolia, la-ti-*fo*-lee-a. Broad-leaved.
nitida, ni-ti-da. Glossy.

Hypsela, hip-*see*-la. *Campanulaceae.*
From Gk. *hypselos* (high). Creeping,
perennial herbs.
reniformis, ree-ni-*form*-is. Kidney-
shaped.

Hyssopus, hi-*sop*-us. *Labiatae.* Origin
unknown. Deciduous or semi-evergreen shrubs.
officinalis, o-fi-si-*na*-lis. Sold in
shops. Hyssop.

I

Iberis, i-*be*-ris. *Cruciferae*. From Gk. *iberis*, (Iberia). Annual herbs and evergreen sub-shrubs. Candytuft.

  amara, a-*ma*-ra. Bitter. Rocket Candytuft.

  gibraltarica, jib-rol-*ta*-ri-ca. Of Gibraltar.

  saxatilis, sax-*a*-ti-lis. Growing on rocks.

  sempervirens, sem-per-*vi*-renz. Evergreen.

  umbellata, um-bel-*a*-ta. With flowers in umbels. Common Candytuft.

Iberis amara

Idesia, i-*deez*-ee-a. *Flacourtiaceae*. After E. I. Ides.

  polycarpa, pol-i-*kar*-pa. With many fruits.

Ilex, *ie*-lex. *Aquifoliaceae*. From the L. name *ilex* (evergreen oak). Evergreen trees and shrubs. Holly.

  x *altaclerensis,* al-ta-kle-*ren*-sis. Of Highclere (Alta Clera).

  aquifolium, a-kwi-*fo*-lee-um. The L. name. Common Holly.

  camelliifolia, ka-mel-ee-i-*fo*-lee-a. With *Camellia*-like leaves.

  cornuta, kor-*new*-ta. Horned (leaf spines). Horned Holly, Chinese Holly.

  crenata, kre-*na*-ta. With rounded teeth (leaves). Box-leaved Holly.

  hodginsii, ho-*jinz*-ee-ee. After Edward Hodgins.

  lawsoniana, law-son-ee-*a*-na. After the Lawson nursery.

  pernyi, per-nee-ee. After the Abbé Perny.

  serrata, se-*ra*-ta. Saw-toothed. Japanese Winterberry.

Ilex aquifolium

Illicium, il-*lis*-ee-um. *Illiciaceae*. From L. *illicio* (attract), after the fragrance. Evergreen trees and shrubs.

  anisatum, an-i-*sa*-tum. Anise-scented (leaves).

  floridanum, flo-ri-*da*-num. Of Florida.

Impatiens, im-*pat*-ee-enz. *Balsaminaceae.* L. *impatiens* (impatient). The ripe seed pods explode when touched. Hardy and tender, annual and perennial herbs. Busy Lizzie.

balfourii, bal-*for*-ree-ee. After Sir Isaac Balfour. Orange Balsam.

balsamina, bal-sa-*meen*-a. Bearing Balsam. Garden Balsam.

capensis, ka-*pen*-sis. Of the Cape of Good Hope. Orange Balsam.

glandulifera, gland-ew-*li*-fe-ra. Gland-bearing. Himalayan Balsam, Policeman's Helmet.

noli-tangere, no-lee-tang-*ge*-ree. Do not touch me (pods). Touch-me-not.

repens, ree-penz. Creeping.

walleriana, wo-la-ree-*a*-na. After the Rev. Horace Waller. Busy Lizzie, Patience Plant.

Impatiens noli-tangere

Incarvillea, in-kar-*vil*-ee-a. *Bignoniaceae.* After Pierre d'Incarville. Perennial herbs.

compacta, com-*pak*-ta. Compact.

delavayi, del-a-*vay*-ee. After Delavay.

mairei, mair-ee-ee. After Edouard Maire.

grandiflora, grand-i-*flo*-ra. Large-flowered.

olgae, ol-gee. After Olga Fedtschenko.

sinensis, sin-*en*-sis. Of China.

Indigofera, in-di-*go*-fe-ra. *Leguminosae.* From indigo and L. *fero* (bear). Indigo is produced from *I. tinctoria.* Deciduous, perennial shrubs.

decora, de-*ko*-ra. Beautiful.

heterantha, he-te-*ranth*-a. With different flowers.

tinctoria, tink-to-*ree*-a. Used in dyeing.

Inula, *in*-ew-la. *Compositae.* The L. for *I. helenium.* Perennial herbs.

acaulis, a-*kaw*-lis. Stemless.

barbata, bar-*ba*-ta. Bearded.

ensifolia, en-si-*fo*-lee-a. With sword-shaped leaves.

helenium, he-*len*-ee-um. *Helenium*-like. Elecampne.

hookeri, huk-a-ree. After Sir Joseph Hooker.

magnifica, mag-*ni*-fi-ka. Splendid.

Inula Helenium

oculus-christi, ok-ew-lus-*kris*-tee.
Eye of Christ.
orientalis, o-ree-en-*ta*-lis. Eastern.

Iochroma, i-o-*kro*-ma. *Solanaceae.*
From Gk. *ion* (violet) and *chroma*
(colour), after the flowers. Tender
shrubs.
  coccineum, kok-*kin*-ee-um. Scarlet.
  cyaneum, sie-*an*-ee-um. Blue.
  fuchsioides, few-she-*oi*-deez.
  Fuchsia-like.
  grandiflorum, grand-i-*flo*-rum.
  Large-flowered.

Ionopsidium, i-on-op-*sid*-ee-um.
Cruciferae. From Gk. *ion* (violet) and
-opsis (resemblance). Annual herb.
  acaule, a-*kaw*-lee. Stemless. Violet
  Cress.

Ipheion, i-*fee*-on. *Liliaceae.* Bulbous
herb.
  uniflorum, ew-ni-*flo*-rum. With one
  flower.

Ipomoea, i-pom-*ee*-a.
Convolvulaceae. From Gk. *ips* (worm)
and *homoios* (resembling). Tender
evergreen annual and perennial
climbers. Morning Glory.
  alba, al-ba. White. Moonflower
  batatas, ba-*ta*-tas. Haitian name.
  Sweet Potato Vine.
  coccinea, kok-*kin*-ee-a. Scarlet. Red
  Morning Glory.
  hederacea, he-de-*ra*-see-a. *Hedera*-
  like (leaves).
  horsfalliae, hors-*fal*-ee-ee. After Mrs
  Charles Horsfall.
  purpurea, pur-*pur*-ree-a. Purple.
  Common Morning Glory.
  quamoclit, kwa-mo-klit. Mexican
  name. Cypress Vine, Indian Pink.
  tricolor, tri-ko-lor. Three-coloured.
  violacea, vie-o-*la*-see-a. Violet.

Ipomopsis, i-pom-*op*-sis.
Polemoniaceae. From Gk. *ips* (worm)
and *-opsis (*resemblance). Annual
herbs.
  aggregata, ag-re-*ga*-ta. Clustered.
  Skyrocket.
  rubra, rub-ra. Red. Standing
  Cypress.

Iresine, i-res-*ee*-nee. *Amaranthaceae.*
From Gk. *eiros* (woolly). Tender,
perennial herbs.
  herbstii, herb-stee-ee. After
  Hermann Herbst. Beef Plant.
  lindenii, lin-*den*-ee-ee. After Linden.
  Blood Leaf.

Iris, *ie*-ris. *Iridaceae.* From Gk. *iris*
(rainbow). Rhizomatous or bulbous
perennial herbs.
  aurea, aw-ree-a. Golden.
  bakeriana, bay-ka-ree-*a*-na. After J.
  G. Baker.
  bucharica, bew-*ka*-ri-ka. Of
  Bokhara.
  bulleyana, bul-ee-*a*-na. After Arthur
  Bulley.
  clarkei, klark-ee-ee. After Charles
  Clarke.
  confusa, kon-*few*-sa. Confused
  cristata, kris-*ta*-ta. Crested.
  cuprea, kew-pree-a. Coppery.
  danfordiae, dan-*ford*-ee-ee. After
  Mrs C. G. Danford.
  douglasiana, dug-las-ee-*a*-na. After
  David Douglas.
  ensata, en-*sa*-ta. Sword-like
  (leaves).
  foetidissima, fee-ti-*dis*-i-ma. Very
  foetid. Stinking Gladwyn.
  forrestii, fo-*rest*-ee-ee. After Forrest.
  fulva, ful-va. Tawny.
  gatesii, gayts-ee-ee. After the Rev. F.
  S. Gates.
  germanica, jer-*man*-i-ka. Of
  Germany.

Iris foetidissima

graebneriana, grayb-na-ree-*a*-na. After Karl Graebner.
graminea, gra-*min*-ee-a. Grass-like.
hoogiana, hoog-ee-*a*-na. After John Hoog.
innominata, in-nom-i-*na*-ta. Nameless.
japonica, ja-*pon*-i-ka. Of Japan.
laevigata, lee-vi-*ga*-ta. Smooth (leaves).
latifolia, la-ti-*fo*-lee-a. Broad-leaved. English Iris.
lutescens, loo-*tes*-enz. Yellowish (flowers).
missouriensis, mi-sur-ree-*en*-sis. Of the Missouri River.
orchioides, or-kee-*oi*-deez. Orchid-like.
orientalis, o-ree-en-*ta*-lis. Eastern.
pallida, pa-li-da. Pale.
pseudacorus, sood-*a*-ko-rus. False *Acorus.* Yellow Flag.
pumila, *pew*-mi-la. Dwarf.
reticulata, ree-tik-ew-*la*-ta. Netted (bulb).
ruthenica, roo-*then*-i-ka. Of Ruthenia.

setosa, see-*to*-sa. Bristly.
sibirica, si-*bi*-ri-ka. Of Siberia.
spuria, *spew*-ree-a. False.
stolonifera, sto-lon-*i*-fe-ra. Bearing stolons.
stylosa, sti-*lo*-sa. With a prominent style.
susiana, soo-see-*a*-na. Of Susa. Mourning Iris.
tectorum, tek-*to*-rum. Growing on roofs. Roof Iris.
tenax, *ten*-ax. Tough (leaves).
tingitana, tin-ji-*ta*-na. Of Tingi.
tuberosa, tew-be-*ro*-sa. Tuberous.
unguicularis, un-gwik-ew-*la*-ris. Clawed. Algerian Iris.
verna, *ver*-na. Of Spring.
versicolor, ver-*si*-ko-lor. Variously coloured.
xiphium, *zi*-fee-um. Swordlike. Spanish Iris.

Isatis, *i*-sa-tis. *Cruciferae.* Gk. name. Biennial herb.
glauca, *glaw*-ka. Glaucous.
tinctoria, tink-*to*-ree-a. Used in dyeing. Dyer's Woad.

Isatis tinctoria

Itea, *i*-tee-a. *Grossulariaceae*. Gk. for willow. Deciduous or evergreen shrubs.
   ilicifolia, i-lis-i-*fo*-lee-a. *Ilex*-leaved.
   virginica, vir-*jin*-i-ka. Of Virginia. Virginia Willow.

Ixia, *ix*-ee-a. *Iridaceae*. From Gk. *ixia* (bird lime), after the sticky sap. Semi-hardy, cormous perennial. Corn Lily.

maculata, mak-ew-*la*-ta. Spotted.
viridiflora, vi-ri-di-*flo*-ra. With green flowers.

Ixora, ix-*o*-ra. *Rubiaceae*. After *Iswara,* a Malabar deity. Tender, evergreen shrubs.
   chinensis, chin-*en*-sis. Of China.
   coccinea, kok-*kin*-ee-a. Scarlet. Flame of the Woods.

J

Jacaranda, jak-a-*rand*-a.
Bignoniaceae. From the Brazilian Indian name. Tender, deciduous or evergreen trees.
  arborea, ar-*bo*-ree-a. Tree-like.
  mimosifolia, mee-mo-si-*fo*-lee-a. *Mimosa*-leaved.
  ovalifolia, o-va-li-*fo*-lee-a. With oval leaves.

Jasione, ja-see-*o*-nee.
Campanulaceae. Gk. name. Perennial herbs. Sheep's Bit.
  crispa, kris-pa. Finely waved.
  laevis, lee-vis. Smooth.
  montana, mon-*ta*-na. Of mountains.

  Jasione montana

Jasminum, jas-*min*-um. *Oleaceae*.
From *yasmin,* the Arabic name. Deciduous or evergreen shrubs and climbers. Jasmine, Jessamine.
  azoricum, a-*zo*-ri-kum. Of the Azores.
  beesianum, beez-ee-*a*-num. After Bees nursery.

floridum, flo-ri-dum. Flowering.
grandiflorum, gran-di-*flo*-rum. With large flowers.
humile, hum-i-lee. Low growing. Italian Yellow Jasmine.
mesnyi, mez-nee-ee. After William Mesny. Primrose Jasmine.
nudiflorum, new-di-*flo*-rum.With flowers appearing before the leaves. Winter Jasmine.
officinale, o-fis-i-*na*-lee. Common Jasmine, Jessamine.
parkeri, park-a-ree. After R. N. Parker.
polyanthum, po-lee-*anth*-um. Many-flowered.
primulinum, prim-ew-*leen*-um. Primrose-coloured. Primrose Jasmine.
revolutum, re-vo-*loo*-tum. Rolled back (leaf margins).

Jeffersonia, jef-er-*son*-ee-a.
Berberidaceae. After President Thomas Jefferson. Perennial herbs.
  diphylla, di-*fil*-la. Two-leaved. Rheumatism Root.
  dubia, dub-ee-a. Doubtful.

Jovibarba, jov-i-*bar*-ba.
Crassulaceae. From L. *Jovis* (Jupiter) and *barba* (beard), after the fringed petals. Succulent, evergreen perennial herbs.
  hirta, hir-ta. Hairy.
  sobolifera, so-bo-*li*-fe-ra. With creeping root stems.

Juglans, *joo*-glans. *Juglandaceae*.
From L. *jovis* (Jupiter) and *glans* (nut). Deciduous trees. Walnut.

ailantifolia, ie-lan-ti-*fo*-lee-a.
Ailanthus-leaved. Japanese Walnut.
californica, ka-li-*forn*-i-ka. Of
California. California Walnut.
cinerea, sin-*e*-ree-a. Grey (bark).
White Walnut.
major, ma-jor. Larger.
microcarpa, mik-ro-*kar*-pa. With
small fruit. Texan Walnut.
nigra, nig-ra. Black (bark). Black
Walnut.
regia, ree-jee-a. Regal. English
Walnut.
rupestris, roo-*pes*-tris. Growing on
rocks.

Juncus, *jun*-kus. *Juncaceae.* From L.
jungo (bind). Perennial herbs.
   effusus, e-*few*-sus. Loosely spreading. Common Rush.

Juniperus, joo-*ni*-pe-rus.
Cupressaceae. L. name. Evergreen
conifers. Juniper.
   californica, ka-li-*forn*-i-ka. Of
California.
   chinensis, chin-*en*-sis. Of China.
Chinese Juniper.
   communis, kom-*ew*-nis. Common.
Common Juniper.
   conferta, kon-*fer*-ta. Crowded. Shore
Juniper.
   coxii, kox-ee-ee. After E. H. M. Cox.
   drupacea, droo-*pa*-see-a. With
fleshy fruit. Syrian Juniper.
   horizontalis, ho-ri-zon-*ta*-lis.
Horizontal. Creeping Juniper.
   procumbens, pro-*kum*-benz. Prostrate.
   recurva, re-*kur*-va. Curved downwards (shoots).

Juniperus communis

   rigida, ri-ji-da. Rigid (leaves).
Temple Juniper.
   sabina, sa-*been*-a. L. name. Savin.
   sargentii, sar-*jent*-ee-ee. After
Sargent.
   scopulorum, skop-ew-*lo*-rum.
Growing on cliffs. Rocky Mountain
Juniper.
   squamata, skwa-*ma*-ta. Scaly (bark).
Flaky Juniper.
   virginiana, vir-jin-ee-*a*-na. Of
Virginia. Eastern Red Cedar.

Justicia, jus-*tis*-ee-a. *Acanthaceae.*
After James Justice. Tender, evergreen
perennial shrubs. Water Willow.
   brandegeana, brand-ee-jee-*a*-na.
After Brandegee. Shrimp Plant.
   carnea, kar-nee-a. Flesh-coloured
(flowers). Flamingo Plant, Brazilian
Plume.
   coccinea, kok-*kin*-ee-a. Scarlet
   floribunda, flo-ri-*bun*-da. Profusely
flowering.
   rizzinii, ritz-*in*-ee-ee. After Carlos
Rizzini.

K

Kalanchoe, ka-*lan*-ko-ee.
Crassulaceae. From the Chinese.
Tender, perennial succulents.
   beharensis, bee-ha-*ren*-sis. Of
   Behara.
   bentii, bent-ee-ee. After Theodore
   Bent.
   blossfeldiana, bloss-feld-ee-*a*-na.
   After Robert Blossfeld. Flaming Katy.
   crenata, kree-*na*-ta. With rounded
   teeth (leaves).
   daigremontiana, day-gre-mont-ee-*a*-
   na. After Mme. and M. Daigremont.
   Devil's Backbone, Mother of
   Thousands.
   fedtschenkoi, fet-*shenk*-o-ee. After
   Boris Fedtschenko.
   flammea, flam-ee-a. Flame-coloured
   (flowers).
   marmorata, mar-mo-*ra*-ta. Marbled
   (leaves). Penwiper.
   pinnata, pin-*a*-ta. Pinnate.
   pumila, pew-mi-la. Dwarf.
   tomentosa, to-men-*to*-sa. Hairy.
   Panda Plant.
   tubiflora, tew-bi-*flo*-ra. With tubular
   flowers.
   uniflora, ew-ni-*flo*-ra. With one
   flower.

Kalmia, *kal*-mee-a. *Ericaceae*. After
Pehr Kalm. Evergreen shrubs.
   angustifolia, an-gust-i-*fo*-lee-a.
   Narrow-leaved. Sheep Laurel.
   latifolia, la-ti-*fo*-lee-a. Broad-leaved,
   Mountain Laurel.
   polifolia, pol-i-*fo*-lee-a. With leaves
   like *Teucrium polium*. Bog Laurel.

Kalmiopsis, kal-mee-*op*-sis.
Ericaceae. From *Kalmia* and Gk.

-opsis (resemblance). Evergreen shrub.
   leachiana, leech-ee-*a*-na. After Mr
   and Mrs Leach.

Kalopanax, kal-o-*pan*-ax. *Araliaceae*.
From Gk. *kalos* (beautiful) and *Panax*.
Deciduous tree.
   pictus, pik-tus. Painted.
   septemlobus, sep-*tem*-lo-bus. With
   seven lobes. Tree Aralia.

Kerria, *ke*-ree-a. *Rosaceae*. After
William Kerr. Deciduous shrub.
   japonica, ja-*pon*-i-ka. Of Japan.

Kirengeshoma, ki-reng-ge-*sho*-ma.
Hydrangeaceae. The Japanese name.
Herbaceous perennial.
   palmata, pal-*ma*-ta. Lobed like a
   hand (leaves).

Kniphofia, nee-*fof*-ee-a. *Liliaceae*.
After Johann Kniphof. Hardy and
semi-hardy perennial herbs. Red Hot
Poker.
   angustifolia, an-gust-i-*fo*-lee-a.
   Narrow-leaved.
   caulescens, kaw-*les*-enz. With a
   stem.
   citrina, si-*tree*-na. Lemon-yellow
   foliosa, fo-lee-*o*-sa. Leafy.
   galpinii, gal-pin-ee-ee. After Ernst
   Galpin.
   gracilis, gra-si-lis. Graceful.
   laxiflora, lax-i-*flo*-ra. With drooping
   flowers.
   multiflora, mul-ti-*flo*-ra. Many-flowered.
   nelsonii, nel-*son*-ee-ee. After
   William Nelson.
   northiae, north-ee-ee. After Miss

Marianne North.
splendida, splen-di-da. Splendid.
uvaria, oo-*va*-ree-a. Like a bunch of
grapes.

Kochia, *kok*-ee-a. *Chenopodiaceae.*
After Wilhelm Daniel Josef Koch.
Annual and perennial herb.
scoparia, sko-*pa*-ree-a. Broom-like.
Summer Cypress, Burning Bush.
trichophylla, tri-ko-*fil*-la. With hair-
like leaves.

Koelreuteria, kurl-roy-*te*-ree-a.
Sapindaceae. After Joseph Gottlieb
Koelreuter. Deciduous tree.
elegans, e-le-ganz. Elegant. Chinese
Rain Tree.
paniculata, pa-nik-ew-*la*-ta. With
flowers in panicles. Golden Rain Tree.

Kohleria, ko-*le*-ree-a. *Gesneriaceae.*
After Michael Kohler. Tender perennial herbs and shrubs.
amabilis, a-*ma*-bi-lis. Beautiful.
bella, be-la. Pretty.
bogotensis, bo-go-*ten*-sis. Of
Bogota.
digitaliflora, di-ji-*ta*-li-flo-ra.
Digitalis-like flowers.
eriantha, e-ree-*anth*-a. With woolly
flowers.
hirsuta, hir-*soo*-ta. Hairy.
spicata, spee-*ka*-ta. With flowers in
spikes.

Kolkwitzia, kol-*kwitz*-ee-a.
Caprifoliaceae. After Richard
Kolkwitz. Deciduous shrub. Beauty
Bush.
amabilis, a-*ma*-bi-lis. Beautiful.

L

Laburnum, la-*burn*-um.
Leguminosae. L. name. Deciduous trees.
   alpinum, al-*pie*-num. Alpine. Scotch Laburnum.
   anagyroides, an-a-gi-*roi*-deez. *Anagyris*-like. Common Laburnum. Golden Chain.
   x *watereri, war*-ta-ra-ree. After the Waterer nursery.

Lachenalia, la-shen-*al*-ee-a. *Liliaceae.* After Werner de La Chenal. Tender, bulbous herbs. Cape Cowslip.
   aloides, a-lo-*ee*-deez. *Aloe*-like.
   bulbifera, bul-*bi*-fe-ra. Bulb-bearing.
   glaucina, glow-*seen*-a. Glaucous.
   mutabilis, mew-*ta*-bi-lis. Changeable (flower colour).
   orchioides, or-kee-*oi*-deez. Orchid-like.

Lactuca alpina

Lactuca, lak-*too*-ka. *Compositae.* From L. *lac* (milk), after the white sap. Annual herb.
   alpina, al-*pie*-na. Alpine. Mountain Sow Thistle.
   perennis, pe-*re*-nis. Perennial. Blue Lettuce.
   plumieri, ploo-mee-*e*-ree. After Charles Plumier.
   sativa, sa-*tee*-va. Cultivated. Garden Lettuce.

Laelia, *lee*-lee-a. *Orchidaceae.* After Laelia, a vestal virgin. Greenhouse orchids.
   cinnabarina, sin-a-ba-*reen*-a. Cinnabar-red.
   grandis, grand-is. Large (flowers).
   pumila, pew-mi-la. Dwarf.
   purpurata, pur-pur-*ra*-ta. Purple (flowers).
   speciosa, spes-ee-*o*-sa. Showy.
   tenebrosa, ten-e-*bro*-sa. Growing in shady places.
   xanthina, zanth-*ee*-na. Yellow (flowers).

Lagarosiphon, la-ga-ro-*see*-fon. *Hydrocharitaceae.* From Gk. *lagaros* (narrow) and *siphon* (tube). Perennial aquatic herb. Curly Water Thyme.
   major, ma-jor. Larger.

Lagerstroemia, la-ger-*strurm*-ee-a. *Lythraceae.* After Magnus von Lagerstrom. Semi-hardy, deciduous trees.
   indica, in-di-ka. Of India. Crape Myrtle.
   speciosa, spes-ee-*o*-sa. Showy. Queen's Crape Myrtle.

Lagurus, la-*gew*-rus. *Gramineae.*
From Gk. *lagos* (hare) and *oura* (tail),
after the inflorescence. Annual grass.
  ovatus, o-*va*-tus. Ovate (inflorescence). Hare's Tail Grass.

Lagurus ovatus

Lamium, *la*-mee-um. *Labiatae.* L.
name. Perennial herbs. Deadnettle.
  album, al-bum. White. White
Deadnettle.

Lamium maculatum

  galeobdolon, ga-lee-*ob*-do-lon.
Stinking.
  garganicum, gar-*ga*-ni-kum. Of
Monte Gargano.
  maculatum, mak-ew-*la*-tum. Spotted
(leaves).
  orvala, or-*va*-la. Sage-like.

Lampranthus, lam-*pranth*-us.
Aizoaceae. From Gk. *lampros* (shining) and *anthos* (flower). Tender, succulent sub-shrubs.
  aurantiacus, aw-ran-tee-*a*-kus.
Orange.
  aureus, aw-ree-us. Golden.
  blandus, bland-us. Mild.
  brownii, brown-ee-ee. After Robert
Brown.
  coccineus, kok-*kin*-ee-us. Scarlet.
  conspicuus, kon-*spik*-ew-us.
Conspicuous.
  haworthii, hay-*werth*-ee-ee. After
Haworth.
  multiradiatus, mul-ti-ra-dee-*a*-tus.
With many rays.
  spectabilis, spek-*ta*-bi-lis.
Spectacular.

Lantana, lan-*ta*-na. *Verbenaceae.* L.
name for *Viburnum,* after its similar
inflorescence. Tender evergreen
shrubs.
  camara, ka-*ma*-ra. After Carama.
  montevidensis, mon-tee-vid-*en*-sis.
Of Montevideo.

Lapageria, la-paj-*er*-ee-a.
Philesiaceae. After Josephine de la
Pagerie, wife of Napoleon. Evergreen
climbers.
  rosea, ros-ee-a. Rose-coloured.

Larix, *la*-rix. *Pinaceae.* L. name.
Deciduous conifers. Larch.
  decidua, de-*sid*-ew-a. Deciduous.
European Larch.

kaempferi, kemp-fa-ree. After
Engelbert Kaempfer. Japanese
Larch.

Lathyrus tuberosus

Lathyrus, *la*-thi-rus. *Leguminosae.*
Gk. for the pea. Annual and perennial
herbs.
   grandiflorus, grand-i-*flo*-rus. Large-
flowered. Everlasting Pea.
   latifolius, la-ti-*fo*-lee-us. Broad-
leaved.
   odoratus, o-do-*ra*-tus. Scented.
Sweet Pea.
   rotundifolius, ro-tund-i-*fo*-lee-us.
With round leaves. Persian
Everlasting Pea.
   splendens, splen-denz. Splendid.
   sylvestris, sil-*ves*-tris. Of woods.
Flat Pea.
   tuberosus, tew-be-*ro*-sus. Tuberous.
Earth Chestnut.
   vernus, ver-nus. Of spring (flowering). Europe.

Laurus, *low*-rus. *Lauraceae.* L. for *L.
nobilis.* Evergreen trees or shrubs.
   azorica, a-*zo*-ri-ka. Of the Azore.
Canary Laurel.
   nobilis, no-bi-lis. Noble. Bay Laurel.

Lavandula, la-*van*-dew-la. *Labiatae.*
From L. *lavo* (wash). Evergreen
shrubs. Lavender.
   angustifolia, an-gust-i-*fo*-lee-a.
Narrow-leaved. English Lavender.
   dentata, den-*ta*-ta. Toothed. French
Lavender.
   lanata, la-*na*-ta. Woolly.

Lavatera, la-va-*te*-ra. *Malvaceae.*
After the Lavater brothers. Annual
herbs and shrubs.
   arborea, ar-*bo*-ree-a. Tree-like. Tree
Mallow.
   maritima, ma-*ri*-ti-ma. Growing near
the sea.
   occidentalis, ok-si-den-*ta*-lis.
Western.
   olbia, ol-bee-a. After Olbia, France.
Tree Lavatera.
   trimestris, trim-*es*-tris. Of three
months (flowering time).

Lavatera arborea

Layia, *lay*-ee-a. *Compositae.* After G.
Tradescant Lay. Annual herb.
   platyglossa, plat-ee-*glos*-a. Broad-
tongued. Tidy Tips.

Ledebouria, led-de-*bour*-ree-a.
Liliaceae, After Carl Friedrich von
Ledebour. Tender, bulbous herbs.
  ovalifolia, o-va-li-*fo*-lee-a. With oval
leaves.
  socialis, so-see-*a*-lis. Growing in colonies.

Ledum, *lee*-dum. *Ericaceae.* From
Gk. *ledon.* Evergreen shrubs.
  groenlandicum, green-*land*-i-kum.
Of Greenland. Labrador Tea.
  palustre, pa-*lus*-tree. Growing in
marshes. Wild Rosemary.

Leiophyllum, lee-o-*fil*-lum.
Ericacecae. From Gk. *leios* (smooth)
and *phyllon* (leaf), after the glossy
leaves. Evergreen shrub.
  buxifolium, bux-i-*fo*-lee-um. *Buxus*-
leaved. Sand Myrtle.

Leontopodium, lee-on-to-*pod*-ee-um.
Compositae. From Gk. *leon* (lion) and
podion (foot). Perennial herb.
  alpinum, al-*pie*-num. Alpine.
Edelweiss.
  stracheyi, *stray*-kee-ee. After
Lieutenant-General Sir Richard
Strachey.

Leptospermum, lep-to-*sperm*-um.
Myrtaceae. From Gk. *leptos* (slender)
and *sperma* (seed), after the narrow
seeds. Evergreen shrubs.
  flavescens, fla-*ves*-enz. Yellowish.
  humifusum, hum-i-*few*-sum. Prostrate.
  lanigerum, la-*ni*-je-rum. Woolly.
  scoparium, sko-*pa*-ree-um. Broom-
like. Manuka, Tea Tree.

Leucadendron, loo-ka-*den*-dron.
Proteaceae. From Gk. *leukos* (white)
and *dendron* (tree), after the silvery
foliage. Tender tree.
  argenteum, ar-*jen*-tee-um. Silvery.
Silver Tree.

Leucocoryne, loo-ko-*ko*-ri-nee.
Liliaceae. From Gk. *leukos* (white)
and *coryne* (club). Bulbous herbs.
  ixioides, ix-ee-*oi*-deez. *Ixia*-like.
Glory of the Sun.
  odorata, o-do-*ra*-ta. Scented.

Leucojum, loo-*ko*-jum.
Amaryllidaceae. From Gk. *leukon*
(white) and *ion* (violet). Bulbous
herbs. Snowflake.
  aestivum, ees-ti-vum. Of Summer.
Summer Snowflake, Loddon Lily.
  autumnale, aw-tum-*na*-lee. Of
autumn.
  roseum, *ros*-ee-um. Rose-coloured.
  vernum, *ver*-num. Of spring. Spring
Snowflake.

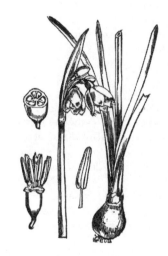

Leucojum aestivum

Leucothoe, loo-*ko*-tho-ee. *Ericaceae.*
After the mythical Leucothoe, daughter of the King of Babylon. Evergreen
and deciduous shrubs.
  fontanesiana, font-a-neez-ee-*a*-na.
After Desfontaines.
  keiskei, *kies*-kee-ee. After Keisuke
Ito.

Lewisia, loo-*is*-ee-a. *Portulacaceae.*
After Captain Meriwether Lewis.
Perennial herbs.
　columbiana, ko-lum-bee-*a*-na. Of
　British Columbia.
　nevadensis, ne-va-*den*-sis. Of the
　Sierra Nevada.
　pygmaea, pig-*mee*-a. Dwarf.
　rediviva, re-di-*vee*-va. Brought back
　to life.
　tweedyi, *twee*-dee-ee. After Tweedy.

Leycesteria, lest-*e*-ree-a.
Caprifoliaceae. Deciduous shrub.
　formosa, for-*mo*-sa. Beautiful.
　Himalaya Honeysuckle.

Liatris, li-*at*-ris. *Compositae.*
Perennial herbs. Gay Feather.
　pycnostachya, pik-no-*stak*-ee-a.
　With dense spikes. Button
　Snakeroot.
　spicata, spee-*ka*-ta. With flowers in
　a spike. Button Snakewort.

Libertia, li-*bert*-ee-a. *Iridaceae.*
After Marie Libert. Perennial herbs.
　grandiflora, grand-i-*flo*-ra. Large-
　flowered.
　ixioides, ix-ee-*oi*-deez. *Ixia*-like.

Ligularia, lig-ew-*la*-ree-a.
Compositae. From L. *ligula* (strap),
after the strap-like florets. Perennial
herbs.
　dentata, den-*ta*-ta. Toothed (leaves).
　Leopard Plant.
　hodgsonii, hoj-*son*-ee-ee. After C. P.
　Hodgson.
　japonica, ja-*pon*-i-ka. Of Japan.
　przewalskii, sha-*val*-skee-ee. After
　Nicolai Przewalski.
　stenocephala, sten-o-*sef*-a-la.
　Narrow-headed.
　tussilaginea, tus-i-la-*gin*-ee-a.
　Tussilago-like.

wilsoniana, wil-son-ee-*a*-na. After
Ernest Wilson. Giant Groundsel.

Ligustrum, li-*gus*-trum, *Oleaceae.* L.
name. Evergreen and semi-evergreen
trees and shrubs. Privet.
　japonicum, ja-*pon*-i-kum. Of Japan.
　lucidum, loo-si-dum. Glossy.
　Chinese Privet.
　obtusifolium, ob-tew-si-*fo*-lee-um.
　Blunt-leaved.
　ovalifolium, o-va-li-*fo*-lee-um. Oval-
　leaved. California Privet.
　sinense, si-*nen*-see. Of China.
　vulgare, vul-*ga*-ree. Common.
　Common Privet.

Ligustrum vulgare

Lilium, *lil*-ee-um. *Liliaceae.* The L.
name. Bulbous perennial herbs. Lily.
　amabile, a-*ma*-bi-lee. Beautiful.
　auratum, aw-*ra*-tum. Marked with
　gold. Mountain Lily.
　bulbiferum, bul-*bi*-fe-rum. Bearing
　bulbs. Fire Lily.
　canadense, kan-a-*den*-see. Of
　Canada. Meadow Lily.
　candidum, *kan*-di-dum. White.

Madonna Lily, White Lily.
cernuum, *sern*-ew-um. Nodding.
Nodding Lily.
chalcedonicum, kal-see-*do*-ni-kum.
Of Chalcedon.
davidii, da-*vid*-ee-ee. After David.
henryi, hen-ree-ee. After Henry.
japonicum, ja-*pon*-i-kum. Of Japan.
Bamboo Lily.
lancifolium, lan-si-*fo*-lee-um. With
lance-shaped leaves. Tiger Lily.
leichtlinii, liekt-*lin*-ee-ee. After Max
Leichtlin.
longiflorum, long-i-*flo*-rum. With
long flowers. Easter Lily.
mackliniae, ma-*klin*-ee-ee. After
Mrs. Macklin.
nepalense, ne-pa-*len*-see. Of Nepal.
pardalinum, par-da-*leen*-um. Spotted
like a leopard. Panther Lily.
pumilum, pew-mi-lum. Dwarf.
pyrenaicum, pi-ren-*ee*-i-kum. Of the
Pyrenees.
regale, ree-*ga*-lee. Regal.
rubellum, rub-*el*-lum. Reddish.
speciosum, spes-ee-*o*-sum. Showy.
superbum, soo-*per*-bum. Superb.
Turk's Cap Lily.
tsingtauense, tsing-tow-*en*-see. Of
Tsingtao.
wallichianum, wo-lik-ee-*a*-num.
After Wallich.

Limnanthes, lim-*nanth*-eez.
Limnanthaceae. From Gk. *limne*
(marsh) and *anthos* (flower). Annual
herbs.
douglasii, dug-*las*-ee-ee. After David
Douglas. Poached Egg Flower.

Limonium, li-*mo*-nee-um.
Plumbaginaceae. From Gk. *leimon*
(meadow). Annual, biennial and
perennial herbs. Sea Lavender.
bellidifolium, bel-i-di-*fo*-lee-um.
With *Bellis*-like leaves.

latifolium, la-ti-*fo*-lee-um. Broad-
leaved.
sinuatum, sin-ew-*a*-tum. Wavy-
edged (leaves).
spicatum, spee-*ka*-tum. With flowers
in spikes.

Linaria, lin-*ar*-ee-a.
Scrophulariaceae. From Gk. *linon*
(flax), after its similar leaves. Annual
and perennial herbs. Toadflax.
alpina, al-*pie*-na. Alpine. Alpine
Toadflax.
dalmatica, dal-*ma*-ti-ka. Of
Dalmatia.
genistifolia, jen-i-sti-*fo*-lee-a.
Genista-leaved.
maroccana, ma-ro-*ka*-na. Of
Morocco. Bunny Rabbits.
purpurea, pur-*pur*-ree-a. Purple.
repens, ree-penz. Creeping. Striped
Toadflax.
reticulata, ree-tik-ew-*la*-ta. Net-
veined (corolla). Purple-net
Toadflax.
triornithophora, tri-or-ni-*tho*-for-a.
Triple-flowered. Three Birds Flying.
vulgaris, vul-*ga*-ris. Common.

Linaria vulgaris

Common Toadflax. Wild
Snapdragon.

Lindera, lin-*de*-ra. *Lauraceae*. After
Johann Linder. Deciduous or evergreen shrubs and trees.
  benzoin, ben-zo-in. From the Arabic
  name. Spice Bush.
  obtusiloba, ob-tew-si-*lo*-ba. Bluntly-
  lobed (leaves).

Linnaea, lin-*ee*-a. *Caprifoliaceae*.
After Linnaeus. Evergreen sub-shrub.
Twin Flower.
  borealis, bo-ree-*a*-lis. Northern.

Linum, *lin*-um. *Linaceae*. L. for flax.
Annual and perennial herbs and
shrubs.
  alpinum, al-*pie*-num. Alpine.
  arboreum, ar-*bor*-ee-um. Tree-like.
  Tree Flax.
  flavum, fla-vum. Yellow. Golden
  Flax.
  grandiflorum, grand-i-*flo*-rum.
  Large-flowered.
  narbonense, nar-bon-*en*-see. Of
  Narbonne.

Linum perenne

perenne, pe-*ren*-ee. Perennial.
salsoloides, sal-so-*loi*-deez. *Salsola*-
like.
usitatissimum, ew-see-ta-*tis*-i-mum.
Most useful. Flax.

Liquidambar, li-kwid-*am*-bar,
Hamamelidaceae. From. L. *liquidus*
(liquid) and *ambar* (amber), after the
bark resin. Deciduous trees.
  formosana, for-mo-*sa*-na. Of
  Formosa.
  styraciflua, sti-ra-*si*-floo-a. Flowing
  with styrax. Sweet Gum.

Liriodendron, li-ri-o-*den*-dron.
Magnoliaceae. From Gk. *leiron* (lily)
and *dendron* (tree). Deciduous trees.
  chinense, chin-*en*-see. Of China.
  Chinese Tulip Tree.
  tulipifera, tew-lip-*i*-fe-ra. Tulip-bear-
  ing. Tulip Tree, Yellow Poplar.

Liriope, li-*ri*-o-pee. *Liliaceae*. After
Liriope, mother of Narcissus.
Evergreen, perennial herbs.
  muscari, mus-*ka*-ree. *Muscari*-like.
  spicata, spee-*ka*-ta. With flowers in
  spikes.

Lithocarpus, lith-o-*kar*-pus.
Fagaceae. From Gk. *lithos* (stone) and
karpos (fruit), after the acorns.
Evergreen trees.
  densiflorus, dens-i-*flo*-rus. Densely
  flowered. Tanbark Oak.
  henryi, hen-ree-ee. After Henry.

Lithodora, lith-o-*do*-ra.
Boraginaceae. From Gk. *lithos* (stone)
and *dorea* (gift). Evergreen shrubs.
  diffusa, di-*few*-sa. Spreading.
  oleifolia, o-lee-i-*fo*-lee-a. With *Olea*-
  like leaves.

Lithops, *lith*-ops. *Aizoaceae*. From

Gk. *lithos* (stone) and *ops* (appearance). Tender, stone-like succulents. Living Stones.
  bella, bel-la. Pretty.
  fulleri, ful-a-ree. After Mr E. R. Fuller.
  karasmontana, ka-ras-mon-*ta*-na. Of the Little Karasberg Mountains.
  lesliei, lez-lee-ee. After Mr T. N. Leslie.
  marmorata, mar-mo-*ra*-ta. Marbled.
  olivacea, o-li-*va*-see-a. Olive green.
  pseudotruncatella, soo-do-trunk-a-*tel*-la. False *L. truncatella.*
  turbiniformis, tur-bin-i-*form*-is. Top-shaped.

Littonia, li-*ton*-ee-a. *Liliaceae.* After Samuel Litton. Tender perennial herbs.
  modesta, mo-*des*-ta. Modest.

Livistona, li-vi-*ston*-a. *Palmae.* After Patrick Murray, Baron of Livingstone. Tender evergreen palms.
  australis, aw-*stra*-lis. Southern. Australian Palm. Cabbage Palm.
  chinensis, chin-*en*-sis. Of China. Chinese Fan Palm.

Lobelia, lo-*bee*-lee-a. *Campanulaceae.* After Mathias de l'Obel. Annual and perennial shrubs.
  cardinalis, kar-di-*na*-lis. Scarlet. Cardinal Flower, Indian Pink.
  erinus, e-ri-nus. Gk. name. Trailing Lobelia.
  fulgens, ful-jenz. Shining.
  siphilitica, si-fi-*li*-ti-ka. After its alleged healing properties. Great Lobelia.
  splendens, splen-denz. Splendid.
  tenuior, ten-*ew*-ee-or. Slender.

Lobivia, lo-*biv*-ee-a. *Cactaceae.* An anagram of Bolivia. Cob Cactus.
  bruchii, bruk-ee-ee. After Bruch.

caespitosa, see-spi-*to*-sa. Tufted.
densispina, dens-i-*speen*-a. Densely spined.
ferox, fe-rox. Spiny.
grandiflora, gran-di-*flo*-ra. With large flowers.
hertichiana, her-trik-ee-*a*-na. After William Hertrich.
pygmaea, pig-*mee*-a. Dwarf.

Lobularia, lob-ew-*la*-ree-a. *Cruciferae.* From L. *lobulus* (small pod), after the fruit. Annual or perennial herbs.
  maritima, ma-*ri*-ti-ma. Growing near the sea. Sweet Alyssum.

Loiseleuria, lwa-ze-*lur*-ree-a. *Ericaceae.* After Jean Loiseleur-Deslongchamps. Prostrate, creeping, evergreen shrub.
  procumbens, pro-*kum*-benz. Prostrate. Alpine Azalea, Mountain Azalea.

Loiseleuria procumbens

Lomatia, lo-*ma*-she-a. *Proteaceae.* From Gk. *loma* (border). The seeds have winged edges. Evergreen shrubs.
  dentata, den-*ta*-ta. Toothed (leaves).
  ferruginea, fe-roo-*jin*-ee-a. Rusty.
  hirsuta, hir-*soo*-ta. Hairy.
  myricoides, mi-ri-*koi*-deez. *Myrica*-like.
  tinctoria, tink-*to*-ree-a. Used in dyeing.

Lonicera caprifolium

Lonicera, lon-i-*se*-ra. *Caprifoliaceae.*
After Adam Lonitzer. Deciduous or
evergreen shrubs and climbers.
Honeysuckle.
  albiflora, al-bi-*flo*-ra. White-flow-
  ered. White Honeysuckle.
  x *americana,* a-me-ri-*ka*-na. Of
  America.
  aureoreticulata, aw-ree-o-re-tik-ew-
  la-ta. Veined with gold (leaves).
  x *brownii, brown*-ee-ee. After
  Brown. Scarlet Trumpet
  Honeysuckle.
  caprifolium, kap-ri-*fo*-lee-um.
  Climbing like a goat.
  etrusca, e-*troos*-ka. Of Tuscany.
  fragrantissima, fra-gran-*tis*-i-ma.
  Very fragrant.
  halliana, hawl-ee-*a*-na. After Dr.
  George Hall.
  henryi, hen-ree-ee. After Henry.
  Giant Honeysuckle.
  hildebrandiana, hil-de-brand-ee-*a*-
  na. After Mr A. H. Hildebrand.
  japonica, ja-*pon*-i-ka. Of Japan.
  Japanese Honeysuckle.
  nitida, ni-ti-da. Glossy (leaves).
  periclymenum, pe-ree-*klim*-en-um.
  Gk. for honeysuckle. Honeysuckle,
Woodbine.
  pileata, pi-lee-*a*-ta. Capped (fruit).
  sempervirens, sem-per-*vi*-renz.
  Evergreen. Trumpet Honeysuckle,
  Coral Honeysuckle.
  serotina, se-*ro*-ti-na. Late flowering.
  standishii, stan-*dish*-ee-ee. After
  John Standish.

Lophophora, lo-fo-*fo*-ra. *Cactaceae.*
From Gk. *lophos* (crest) and *phoreo*
(bear).
  williamsii, wil-*yam*-zee-ee. After
  Williams. Dumpling Cactus, Mescal
  Button, Peyote.

Lotus, *lo*-tus. *Leguminosae.* Gk. name
for a number of plants. Perennial,
semi-evergreen sub-shrubs.
  corniculatus, kor-nik-ew-*la*-tus.
  With small horns. Bird's-foot
  Trefoil.
  peliorhynchus, pel-e-or-*in*-kus.
  Stork's Beak.

Luculia, lu-*kew*-lee-a. *Rubiaceae.*
From *lukuli swa*, a Nepalese name.
Tender, evergreen shrubs.
  grandifolia, grand-i-*fo*-lee-a. Large-
  leaved.
  gratissima, gra-*tis*-i-ma. Very pleasing.

Lunaria, loon-*a*-ree-a. *Cruciferae.*
From L. *luna* (moon), after the pod
shape. Annual and perennial herbs.
  annua, an-ew-a. Annual. Honesty,
  Silver Dollar.
  rediviva, re-di-*veev*-a. Reviving.
  Perennial Honesty.

Lupinus, lu-*pie*-nus. *Leguminosae.*
From L. *lupus* (wolf). Annual, perennial and semi-evergreen shrubs. Lupin.
  arboreus, ar-*bo*-ree-us. Tree-like.
  Tree Lupin.

densiflorus, dens-i-*flo*-rus. Densely-flowered.
hartwegii, hart-*weg*-ee-ee. After Karl Theodore Hartweg.
luteus, *loo*-tee-us. Yellow.
polyphyllus, po-li-*fil*-lus. With many leaves (leaflets).
pubescens, pew-*bes*-enz. Hairy.
subcarnosus, sub-kar-*no*-sus. Somewhat fleshy.
texensis, tex-*en*-sis. Of Texas. Texas Bluebonnet.

Luzula, *luz*-ew-la. *Juncaceae.* From Italian *lucciola* (firefly). Perennial herbs. Wood-rush.
campestris, cam-*pes*-tris. Of fields. Field Wood Rush.
nivea, *ni*-vee-a. Snow-white. Snow Rush.
sylvatica, sil-*va*-ti-ka. Of woods.

Luzula sylvatica

Lycaste, lie-*kas*-tee. *Orchidaceae.* After Lycaste, daughter of Priam, King of Troy. Greenhouse orchids.
aromatica, a-ro-*ma*-ti-ka. Fragrant.
cruenta, kroo-*en*-ta. Blood red.

deppei, *dep*-ee-ee. After Ferdinand Deppe.
gigantea, ji-*gan*-tee-a. Very large.
macrophylla, mak-ro-*fil*-la. Large-leaved.

Lychnis alpina

Lychnis, *lik*-nis. *Caryophyllaceae.* From Gk. *lychnos* (lamp). Perennial herbs. Catchfly.
alpina, al-*pie*-na. Alpine. Alpine Campion.
chalcedonica, kal-see-*don*-i-ka. Of Chalcedon. Maltese Cross.
coeli-rosea, see-le-*ro*-see-a. Rose of Heaven.
coronaria, ko-ro-*na*-ree-a. Used in garlands. Rose Campion.
coronata, ko-ro-*na*-ta. Crowned.
flos-jovis, flos-*jov*-is. Flower of Jupiter.
viscaria, vis-*ka*-ree-a. Sticky (stems). German Catchfly.

Lycium, *lie*-see-um. *Solanaceae.* From the Gk. Deciduous shrub. Oxthorn.
barbarum, *bar*-ba-rum. Foreign. Box Thorn, Common Matrimony Vine.
chinense, chin-*en*-see. Of China.

Lycoris, lie-*ko*-ris. *Amaryllidaceae.* After Lycoris, a Roman actress and mistress of Mark Antony. Tender, bulbous herbs.

incarnata, in-kar-*na*-ta. Flesh pink.
radiata, rad-ee-*a*-ta. Radiating (stamens). Spider Lily.
sanguinea, sang-*gwin*-ee-a. Blood red.
squamigera, skwa-*mi*-je-ra. Scaly. Magic Lily.

Lyonia, lie-*on*-ee-a. *Ericaceae.* After John Lyon. Deciduous and semi-evergreen shrubs.

ligustrina, li-gus-*tree*-na. *Ligustrum*-like. Male Blueberry.
mariana, ma-ree-*a*-na. Of Maryland. Stagger Bush.
ovalifolia, o-va-li-*fo*-lee-a. With oval leaves.

Lyonothamnus, lie-on-o-*tham*-nus. *Rosaceae.* After W. S. Lyon and Gk. *thamnos* (shrub). Evergreen tree.

asplenifolius, a-sple-ni-*fo*-lee-us. With *Asplenium*-like leaves.
floribundus, flo-ri-*bun*-dus. Profusely flowering. Catalina Ironwood.

Lysichiton, li-si-*ki*-ton. Araceae. From Gk. *lysis* (releasing) and *chiton* (cloak). Perennial herbs.

americanus, a-me-ri-*ka*-nus. Of America. Skunk Cabbage.
camtschatcense, kamt-shat-*ken*-see. Of Kamchatka.

Lysimachia, li-si-*mak*-ee-a. *Primulaceae.* After King Lysimachus of Thrace. Perennial herbs. Loosestrife.

clethroides, kleth-*roi*-deez. *Clethra*-like. Gooseneck Loosestrife.
ephemerum, e-*fem*-e-rum. From L. name.
nemorum, ne-mo-rum. Growing in woods. Yellow Pimpernel.
nummularia, num-ew-*la*-ree-a. With coin-shaped leaves. Creeping Jenny, Moneywort.
punctata, punk-*ta*-ta. Dotted.

Lythrum salicaria

Lythrum, *lith*-rum. *Lythraceae.* From Gk. *lythron* (blood), after the colour of the flowers. Perennial herbs.

salicaria, sal-i-*sar*-ee-a. *Salix*-like. Purple Loosestrife, Spiked Loosestrife.
virgatum, vir-*ga*-tum. Twiggy.

M

Maackia, *mark*-ee-a. *Leguminosae.* After Richard Maack. Deciduous trees.
    amurensis, am-ew-*ren*-sis. Of the Amur River region.

Mackaya, mak-*kay*-a. *Acanthaceae.* After Dr J F Mackay. Evergreen shrub.
    bella, bell-a. Pretty.

Macleaya, ma-*klay*-a. *Papaveraceae.* After Alexander Macleay. Perennial herbs. Plume Poppy.
    cordata, kor-*da*-ta. Heart-shaped (leaves).
    microcarpa, mik-ro-*kar*-pa. With small fruits.

Maclura, ma-*kloo*-ra. *Moraceae.* After William Maclure. Deciduous tree.
    pomifera, pom-*i*-fe-ra. Apple-bearing. Osage Orange, Bow Wood.

Magnolia, mag-*nol*-ee-a. *Magnoliaceae.* After Pierre Magnol. Deciduous, semi-evergreen and evergreen, trees and shrubs.
    acuminata, a-kew-min-*a*-ta. With a long point (leaves). Cucumber Tree.
    campbellii, kam-*bel*-ee-ee. After Dr Archibald Campbell.
    delavayi, del-a-*vay*-ee. After Delavay.
    fraseri, fray-za-ree. After John Fraser. Ear-leaved Umbrella Tree.
    grandiflora, grand-i-*flo*-ra. Large-flowered. Southern Magnolia.
    kobus, ko-bus. From *kobushi,* the Japanese name.
    liliiflora, lil-ee-i-*flo*-ra. Lily-like leaves. Woody Orchid.
    macrophylla, mak-ro-*fil*-la. Large-leaved. Umbrella Tree.
    salicifolia, sa-li-si-*fo*-lee-a. *Salix*-leaved. Anise Magnolia.
    sieboldii, see-*bold*-ee-ee. After Siebold.
    sinensis, si-*nen*-sis. Of China.
    stellata, ste-*la*-ta. Star-like. Star Magnolia.
    tripetala, tri-*pe*-ta-la. With three petals. Elkwood.
    virginiana, vir-jin-ee-*a*-na. Of Virginia. Sweet Bay, Swamp Laurel.
    wilsonii, wil-*son*-ee-ee. After Ernest Wilson.

Mahonia, ma-*hon*-ee-a. *Berberidaceae.* After Bernard McMahon. Hardy to semi-hardy evergreen shrubs.
    aquifolium, a-kwi-*fo*-lee-um. L. for holly. Oregon Grape.
    bealei, beel-ee-ee. After T. C. Beale.
    japonica, ja-*pon*-i-ka. Of Japan.

Maianthemum bifolium

lomariifolia, lo-ma-ree-i-*fo*-lee-a.
With *Lomaria*-like leaves.
repens, ree-penz. Creeping.

Maianthemum, my-*anth*-e-mum.
Liliaceae. From Gk. *maios* (May) and
anthemon (blossom). Perennial rhizomatous herbs. May Lily.
   bifolium, bi-*fo*-lee-um. Two leaved.
   False Lily of the Valley.
   canadense, kan-a-*den*-see. Of
   Canada. Two-leaved Solomon's
   Seal.

Malcolmia, mal-*kol*-mee-a.
Cruciferae. After William Malcolm.
Annual herbs.
   littorea, lit-*or*-ee-a. Of the seashore.
   maritima, ma-*ri*-ti-ma. Growing near
   the sea. Virginia Stock. Greece,
   Albania.

Malope, *ma*-lo-pee. *Malvaceae.* From
Gk name for Mallow. Annual herb.
   malacoides, ma-la-*koi*-deez.
   Mallow-like.
   trifida, *tri*-fi-da. Three-lobed
   (leaves).

Malus, *ma*-lus. *Rosaceae.* L. for apple.
Deciduous trees and shrubs. Apple.
   baccata, ba-*ka*-ta. Bearing berries.
   Siberian Crab.
   coronaria, ko-ro-*na*-ree-a. Used in
   garlands.
   floribunda, flo-ri-*bun*-da. Profusely
   flowering. Japanese Crab.
   halliana, hawl-ee-*a*-na. After Dr G.
   R. Hall.
   hupehensis, hew-pee-*hen*-sis. Of
   Hupeh, China.
   purpurea, pur-*pur*-ree-a. Purple
   (foliage).
   sargentii, sar-*jent*-ee-ee. After
   Sargent.
   spectabilis, spek-*ta*-bi-lis.

Spectacular. Asiatic Apple.
tschonoskii, chon-*os*-kee-ee. After
Tschonoski.

Malva, *mal*-va. *Malvaceae.* L. for
mallow. Annual and biennial herbs.
Mallow.
   moschata, mos-*ka*-ta. Musky. Musk
   Mallow.
   neglectus, neg-*lec*-tus. Common
   Mallow.
   sylvestris, sil-*ves*-tris. Of woods. Tall
   Mallow.

Malva moschata

Malvaviscus, mal-va-*vis*-kus.
Malvaceae. From *Malva* and L. *viscus*
(glue). Tender, evergreen trees and
shrubs.
   arboreus, ar-*bo*-ree-us. Tree-like.
   Wax Mallow.

Mammillaria, ma-mil-*lar*-ee-a.
Cactaceae. From L. *mammilla* (nipple).
   bocasana, bo-ka-*sa*-na. Of the Sierra
   de Bocas.
   candida, *kan*-di-da. White (spines).
   Snowball Pincushion.
   densispina, dens-i-*speen*-a. Densely
   spiny.

elegans, e-le-ganz. Elegant.
elongata, e-long-*ga*-ta. Elongated.
hahniana, han-ee-*a*-na. After Hahn.
Old Woman Cactus.
magnimamma, mag-ni-*mam*-a. With
large tubercles.
microhelia, mik-ro-*hee*-lee-a. A
small sun.
prolifera, pro-*li*-fe-ra. Proliferous.
zeilmanniana, ziel-man-ee-*a*-na.
After Zeilmann. Rose Pincushion.

Mandevilla, man-de-*vil*-a.
Apocynaceae. After Henry Mandeville.
Deciduous or evergreen and semi-
evergreen climbers.
boliviensis, bo-liv-ee-*en*-sis. Of
Bolivia. White Dipladenia.
laxa, lax-a. Loose.
splendens, splen-denz. Splendid.
suaveolens, swa-*vee*-o-lenz. Sweetly
scented. Chilean Jasmine.

Mandragora, man-*drag*-o-ra.
Solanaceae. Gk. name. Perennial
herbs.
autumnalis, aw-tum-*na*-lis. Of
autumn.
officinarum, o-fi-si-*na*-rum. Sold in
shops. Mandrake, Devil's Apples.

Manettia, ma-*net*-ee-a. *Rubiaceae.*
After Saveria Manetti. Tender, evergreen climbers.
cordifolia, cor-di-*fol*-ee-a.
Firecracker Vine.
inflata, in-*fla*-ta. Swollen.

Maranta, ma-*ran*-ta. *Marantaceae.*
After Bartolommeo Maranti. Tender,
evergreen, perennial herbs.
arundinacea, a-run-di-*na*-see-a.
Reed-like. Arrowroot, Obedience
Plant.
bicolor, bi-ko-lor. Two-coloured
(leaves).

leuconeura, loo-ko-*newr*-ra. White-
veined (leaves). Prayer Plant, Ten
Commandments.

Margyricapus, mar-ji-ri-*kar*-pus.
Rosaceae. From Gk. *margarites*
(pearl) and *karpos* (fruit). Evergreen
shrubs.
pinnatus, pin-*na*-tus. Pinnate. Pearl
Fruit.
setosus, see-*to*-sus. Bristly.

Masdevallia, mas-de-*va*-lee-a.
Orchidaceae. After Jose Masdevall.
Evergreen, greenhouse orchids.
amabilis, a-*ma*-bi-lis. Beautiful.
bella, bel-la. Pretty.
caudata, kaw-*da*-ta. With a slender
tail (sepals).
chimaera, ki-*mee*-ra. A monster.
coccinea, kok-*kin*-ee-a. Scarlet.
infracta, in-*fract*-a. Curving inward,
tovarensis, to-va-*ren*-sis. Of Tovar,
Venezuela.

Matthiola, mat-ee-*o*-la, *Cruciferae.*
After Piero Mattioli. Annual, biennial
and perennial sub-shrubs. Stock.
bicornis, bi-*kor*-nis. Two-horned
(fruit).
incana, in-*ka*-na. Grey. Brompton
Stock.

Matthiola incana

longipetala, long-i-*pe*-ta-la. Long-petalled. Night-scented Stock.

Maxillaria, max-i-*la*-ree-a. *Orchidaceae.* From L. *maxilla* (jaw). Evergreen greenhouse orchids.
  alba, al-ba. White.
  grandiflora, grand-i-*flo*-ra. Large-flowered.
  sanderiana, san-da-ree-*a*-na. After the Sander nursery.
  tenuifolia, ten-ew-i-*fo*-lee-a. Slender-leaved.
  variabilis, va-ree-*a*-bi-lis. Variable.
  venusta, ven-*us*-ta. Charming.

Mazus, *ma*-zus. *Scrophulariaceae.* From Gk. *mazos* (teat), after the swollen corolla. Creeping, perennial herbs.
  pumilio, pew-*mi*-lee-o. Dwarf.
  reptans, rep-tanz. Creeping.

Meconopsis, mee-ko-*nop*-sis. *Papaveraceae.* From Gk. *mekon* (poppy) and *-opsis* (resemblance). Perennial or monocarpic herbs.
  betonicifolia, be-ton-i-ki-*fo*-lee-a. *Betonica*-leaved. Blue Poppy.
  cambrica, kam-bri-ka. Of Wales. Welsh Poppy.
  grandis, grand-is. Large.
  integrifolia, in-teg-ri-*fo*-lee-a. With entire leaves.
  napaulensis, na-paw-*len*-sis. Of Nepal.
  quintuplinervia, kwin-tup-li-*ner*-vee-a. With five veins (leaves).
  superba, soo-*per*-ba. Superb.

Medinilla, me-di-*ni*-la. *Melastomataceae.* After Jose de Medinilla. Tender, evergreen shrubs and climbers.
  magnifica, mag-*ni*-fi-ka. Magnificent.

Melaleuca, me-la-*loo*-ka. *Myrtaceae.* From Gk. *melas* (black) and *leukos* (white), after the old and new bark. Tender to semi-hardy, evergreen trees and shrubs. Paperbark.
  armillaris, arm-i-*la*-ris. Encircled. Bracelet.
  elliptica, e-*lip*-ti-ka. Elliptic.
  hypericifolia, hi-pe-ree-ki-*fo*-lee-a. *Hypericum*-leaved.
  quinquenervia, kwin-kwee-*ner*-vee-a. With five nerves. Paperbark Tree.
  wilsonii, wil-*son*-ee-ee. After Charles Wilson.

Melianthus, me-lee-*an*-thus. *Melianthaceae.* From Gk. *meli* (honey) and *anthos* (flower). Semi-hardy, evergreen shrubs.
  major, ma-jor. Larger. Honeybush.

Melissa, me-*lis*-a. *Labiatae.* From Gk. name for honeybee. Deciduous, perennial herbs.
  grandiflora, gran-di-*flo*-ra. Large-flowered.
  officinalis, o-fi-si-*na*-lis. Sold in shops. Lemon Balm, Bee balm.

Melittis melissophyllum

Melittis, me-*li*-tis. *Labiatae.* From Gk.
name for honeybee. Perennial herb.
Bastard Balm.
  melissophyllum, me-lis-o-*fil*-lum.
  With *Melissa-l*ike leaves.

Mentha, *men*-tha. *Labiatae.* L. name.
Aromatic, perennial herbs. Mint.
  aquatica, a-*kwa*-ti-ka. Growing in or
  near water. Watermint.
  citrata, si-*tra*-ta. Lemon-scented.
  Eau de Cologne Mint, Lemon Mint.
  longifolia, long-i-*fo*-lee-a. Long-
  leaved. Horse Mint.
  officinalis, o-fi-si-*na*-lis. Sold in
  shops. White Peppermint.
  x *piperita,* pi-pe-*ree*-ta. Like pepper.
  Peppermint.
  pulegium, poo-*leg*-ee-um. L. name
  for flea. Pennyroyal.
  requienii, rek-wee-*en*-ee-ee. After
  Esprit Requien. Corsican Mint.
  x *spicata,* spee-*ka*-ta. With flowers
  in spikes. Spearmint.
  suaveolens, swa-*vee*-o-lenz. Sweetly
  scented. Applemint.

Mentha piperita

Mentzelia, ment-*zel*-ee-a. *Loosaceae.*
After Christian Mentzel. Annual,

perennial and evergreen shrubs.
  lindleyi, lind-lee-ee. After John
  Lindley. Blazing Star.

Menyanthes, mee-nee-*anth*-eez.
Menyanthaceae. From Gk *menanthos*
(moonflower). Deciduous and perennial aquatic herbs.
  trifoliata, tri-fo-lee-*a*-ta. With three
  leaves. Bog Bean, Marsh Trefoil.

Menyanthes trifoliata

Menziesia, men-*zeez*-ee-a. *Ericaceae.*
After Archibald Menzies. Deciduous
shrubs.
  ciliicalyx, si-lee-i-*ka*-lix. With the
  calyx fringed with hairs.
  purpurea, pur-*pur*-ree-a. Purple.

Merendera, me-ren-*de*-ra. *Liliaceae.*
From the Spanish *quita meriendas.*
Cormous herbs.
  bulbocodium, bul-bo-*co*-dee-um.
  With a woody bulb.
  montana, mon-*ta*-na. Of mountains.
  robusta, ro-*bus*-ta. Robust.

Mertensia, mer-*ten*-zee-a.
Boraginaceae. After Franz Karl
Mertens. Perennial herbs.

Mertensia maritima

ciliata, si-lee-*a*-ta. Fringed with hairs (leaves).
echioides, e-kee-*oi*-deez. *Echium*-like.
maritima, ma-*ri*-ti-ma. Growing near the sea.
virginica, vir-*jin*-i-ka. Of Virginia.

Mesembryanthemum, mes-em-bree-*anth*-e-mum. *Aizoaceae.* From Gk *mesembria* (midday) and *anthemon* (flower). Annual and biennial succulents.
crystallinum, kris-ta-*leen*-um. Crystalline (leaves).
tricolor, *tri*-ko-lor. Three-coloured.

Mespilus, *mes*-pi-lus. *Rosaceae.* From Gk. *mesos* (half) and *pilos* (ball), after the fruit shape. Deciduous trees or shrubs.
germanica, ger-*ma*-ni-ka. Of Germany. Medlar.

Metrosideros, me-tro-si-*dee*-ros. *Myrtaceae.* From Gk. *metra* (heartwood) and *sideros* (iron). Evergreen trees, shrubs and climbers.
excelsa, ex-*cel*-sa. Tall. Christmas Tree.

robusta, ro-*bus*-ta. Robust. New Zealand Christmas Tree.

Microbiota, mik-ro-bie-*o*-ta. *Cupressaceae.* From Gk. *micros* (small) and *Biota* (*Thuja*). Prostrate, evergreen conifer.
decussata, dee-kus-*a*-ta. With the leaves in pairs and at right angles to each other.

Milium, *mil*-ee-um. *Gramineae.* L. for millet. Perennial, evergreen grass.
effusum, e-*few*-sum. Spreading. Wood Millet.

Miltonia, mil-*ton*-ee-a. *Orchidaceae.* After Viscount Milton (Charles Fitzwilliam). Greenhouse orchids. Pansy Orchid.
candida, *kan*-di-da. White.
clowesii, *klowz*-ee-ee. After Clowes.
flavescens, fla-*ves*-enz. Yellowish.
spectabilis, spek-*ta*-bi-lis. Spectacular.
warscewiczii, var-sha-*vich*-ee-ee. After Warscewicz.

Mespilus germanica

Milium effusum

Mimosa, mi-*mo*-sa. *Leguminosae.*
From Gk. *mimos* (imitator). The leaves
are sensitive to the touch. Tender
annual or evergreen perennial trees,
shrubs and climbers.
  pudica, pu-*dee*-ka. Shy. Humble
  Plant, Touch-me-not.

Mimulus luteus

sensitiva, sen-sit-*iv*-a. Sensitive to
the touch. Sensitive plant.

Mimulus, *mim*-ew-lus.
Scrophulariaceae. From L. *mimus*
(mimic). The flower looks like the
face of a monkey. Annual, perennial
and evergreen shrubs. Monkey Flower,
Musk.
  aurantiacus, aw-ran-tee-*a*-kus.
  Orange.
  cardinalis, kar-di-*na*-lis. Scarlet.
  Scarlet Monkey Flower.
  guttatus, gu-*ta*-tus. Spotted (flowers). Common Large Monkey
  Flower.
  lewisii, loo-*is*-ee-ee. After Lewis.
  Great Purple Monkey Flower.
  luteus, *loo*-tee-us. Yellow. Monkey
  Musk.
  moschatus, mos-*ka*-tus. Musk
  Flower.
  puniceus, pew-*ni*-see-us. Reddish-
  purple.

Mina, *mie*-na. *Convolvulaceae.* After
Joseph Mina. Tender, semi-evergreen
or deciduous climbers.
  lobata, lo-*ba*-ta. With lobes (leaves).

Mirabilis, mee-*ra*-bi-lis.
Nyctaginaceae. L. for wonderful.
Semi-hardy annual herbs.
  jalapa, ha-*la*-pa. Of Jalapa, Mexico.
  Four o'Clock Plant.
  linearis, lin-ee-*a*-ris. Linear.
  multiflora, mul-ti-*flo*-ra. Many-flowered.

Miscanthus, mis-*kanth*-us.
Gramineae. From Gk. *miskos* (stem)
and *anthos* (flower) the tall spikelets.
Perennial grasses.
  sacchariflorus, sa-ka-ri-*flo*-rus.
  Saccharum-like flowers. Amur
  Silver Grass.

sinensis, si-*nen*-sis. Of China. Eulalia.

Mitchella, mi-*chel*-la. *Rubiaceae.* After Dr John Mitchell. Evergreen sub-shrubs.
repens, ree-penz. Creeping. Partridge Berry, Two-eyed Berry.

Mitella, mi-*tel*-la. *Saxifragaceae.* From L. *mitra* (cap), after the seed pod. Rhizomatous, perennial herbs. Bishop's Cap, Mitrewort.
breweri, broo-a-ree. After Brewer.
caulescens, kaw-*les*-enz. With a stem.
diphylla, di-*fil*-la. Two-leaved.
trifida, tri-fi-da. Three-lobed.

Mitraria, mi-*tra*-ree-a, *Gesneriaceae.* From L. *mitra* (cap), after the bracts. Semi-hardy, evergreen climber.
coccinea, kok-*kin*-ee-a. Scarlet.

Molinia, mo-*leen*-ee-a, *Gramineae.* After Juan Molina. Perennial grass.

Molinia caerulea

caerulea, see-*ru*-lee-a. Dark blue. Purple Moor Grass.

Moltkia, *molt*-kee-a. *Boraginaceae.* After Count Moltke. Deciduous, semi-evergreen perennials and sub-shrubs.
caerulea, see-*ru*-lee-a. Dark blue.
petraea, pe-*tree*-a. Growing on rocks.
suffruticosa, su-froo-ti-*ko*-sa. Sub-shrubby

Moluccella, mo-lu-*kel*-la. *Labiatae.* Origin unknown. Annual and perennial herbs.
laevis, lee-vis. Smooth. Bells of Ireland.
spinosa, spee-*no*-sa. Spiny.

Monarda, mo-*nar*-da. *Labiatae.* After Nicholas Monardes. Aromatic, annual and perennial herbs.
citriodora, sit-ree-o-*do*-ra. Lemon-scented.
didyma, di-di-ma. In pairs (stamens). Bee Balm, Sweet Bergamot.
fistulosa, fist-ew-*lo*-sa. Tubular.
menthifolia, men-thee-*fo*-lee-a. Mint-leaved. Mint-leaved Bergamot.

Monstera, mon-*stee*-ra. *Araceae.* Origin unknown. Tender, evergreen climbers.
deliciosa, dee-li-see-*o*-sa. Delicious. Swiss-cheese Plant.

Moraea, mo-*ree*-a. *Iridaceae.* After Robert More. Cormous perennial herbs.
iridioides, ee-ri-dee-*oi*-dees. *Iris*-like.
spathacea, spa-*tha*-see-a. Spathe-like.
tricuspidata, tri-kus-pi-*da*-ta. Three-pointed.

Morina, mo-*reen*-a. *Morinacaceae.* After Louis Morin. Evergreen, peren-

nial herb.
longifolia, long-i-*fo*-lee-a. Long-leaved. Whorlflower.
persica, per-si-ka. Of Persia.

Morisia, mo-*ris*-ee-a. *Cruciferae.* After Giuseppe Giacinto Moris. Perennial herb.
monanthos, mon-*anth*-os. One-flowered.
hypogaeus, hie-po-*jee*-us. Developing underground.

Morus, *mo*-rus. *Moraceae.* L. name *M. nigra.* Deciduous trees. Mulberry.
alba, al-ba. White. White Mulberry.
nigra, nig-ra. Black. Common Mulberry, Black Mulberry.
rubra, rub-ra. Red. Red Mulberry.

Muehlenbeckia, moo-lan-*bek*-ee-a. *Polygonaceae.* After Dr Gustave Muehlenbeck. Deciduous or evergreen shrubs and climbers.
complexa, kom-*plex*-a. Encircled. Maidenhair Vine, Wire Vine.

Muscari, mus-*ka*-ree. *Liliaceae.* From Gk. *moscos* (musk). Bulbous herbs. Grape Hyacinth.
armeniacum, ar-men-ee-*a*-kum. Of Armenia.
aucheri, aw-ka-ree. After P. M. R. Aucher-Eloy.
azureum, a-*zew*-ree-um. Sky-blue (flowers).
botryoides, bot-ree-*oi*-deez. Like a bunch of grapes.
comosum, ko-*mo*-sum. With a tuft. Tassel Hyacinth.
latifolium, la-ti-*fo*-lee-um. Broad-leaved.
macrocarpum, mak-ro-*kar*-pum. With large fruit. Greece,
moschatum, mos-*ka*-tum. Musk-scented. Musk Hyacinth.

Muscari racemosum

neglectum, ne-*glek*-tum. Overlooked.
paradoxum, pa-ra-*dox*-um. Unusual.
racemosum, ra-see-*mo*-sum. With flowers in racemes.
tubergenianum, tew-ber-gen-ee-*a*-num. After van Tubergen.

Mutisia, mew-*tis*-ee-a. *Compositae.* After Celestino Mutis. Semi-hardy, evergreen climbers. Climbing Gazania.
clematis, kle-ma-tis. Clematis-like.
decurrens, dee-*ku*-renz. The leaf base merges with the stem.
oligodon, o-*li*-go-don. With few teeth.

Myosotidium, mee-os-o-*tid*-ee-um. *Boraginaceae.* From Gr. *Myosotis* and *eidos* (appearance). Semi-hardy, evergreen, perennial herb.
hortensia, hor-*tens*-ee-a. Of gardens. Chatham Island Forget-me-not.

Myosotis, mie-os-*o*-tis. *Boraginaceae.* From L. *mus* (mouse) and Gk. *otos* (ear), after the leaves. Annual, biennial

and perennial herbs. Forget-me-not,
Scorpion Grass.

alpestris, al-*pes*-tris. Of lower
mountains.

alpina, al-*pie*-na. Alpine.

australis, aw-*stra*-lis. Southern.

caespitosa, see-spi-*to*-sa. Tufted.

laxa, lax-a. Loose.

macrantha, ma-*kranth*-a. Large-
flowered.

palustris, pa-*lus*-tris. Growing in bogs.

scorpioides, skor-pee-*oi*-deez. Like a
scorpion.

sylvatica, sil-*va*-ti-ka. Of woods.
Garden Forget-me-not.

uniflora, ew-ni-*flo*-ra. With one
flower.

Myrica, mi-*ree*-ka. *Myricaceae.* From
Gk. *myrike* (Tamarisk). Deciduous and
evergreen shrubs.

californica, kal-i-*forn*-i-ka. Of
California. Californian Bayberry.

cerifera, see-*ri*-fe-ra. Wax-bearing
(fruit). Wax Myrtle.

gale, ga-lee. From Old English.
Sweet Gale, Bog Myrtle.

pensylvanica, pen-sil-*van*-i-ka. Of
Pennsylvania. Bayberry.

Myriophyllum, mi-ree-o-*fil*-lum.
Haloragidaceae. From Gk. *myrios*
(many) and *phyllon* (leaf). Deciduous
and perennial aquatic herbs. Water
Milfoil.

aquaticum, a-*kwa*-ti-kum. Growing
in water. Diamond Milfoil.

verticillatum, ver-ti-si-*la*-tum.
Whorled (leaves). Myriad Leaf.

Myrrhis, *mi*-ris. *Umbelliferae.* Gk. for
Myrrh. Perennial herb.

odorata, o-do-*ra*-ta. Scented. Sweet

Myrrhis odorata

Cicely, Garden Myrrh.

Myrsine, *mur*-si-nee. *Myrsinaceae.*
Gk. for myrtle. Evergreen shrubs and
trees.

africana, af-ri-*ka*-na. African.
African Boxwood, Cape Myrtle.

Myrtillocactus, mur-ti-lo-*kak*-tus.
Cactaceae. From L. *myrtillus* (small
myrtle) and *Cactus,* after the myrtle-
like fruits.

geometrizans, gee-o-*met*-ri-zanz.
Regularly marked.

Myrtus, *mur*-tus. *Myrtaceae.* From
Gk. *Murtos* (Myrtle). Evergreen
shrubs. Myrtle.

apiculata, a-pik-ew-*la*-ta. With an
abrupt, short point.

bullata, bu-*la*-ta. With puckered
leaves.

communis, kom-*ew*-nis. Common.
Myrtle.

luma, loo-ma. The Chilean name.

nummularia, num-ew-*la*-ree-a. With
coin-shaped leaves.

ugni, un-yee. The Chilean name.

N

Nandina, nan-*deen*-a. *Berberidaceae.*
From the Japanese *nandin.* Evergreen
shrub.
  domestica, do-*mes*-ti-ka. Cultivated.
  Heavenly Bamboo.

Narcissus, nar-*sis*-us. *Amaryllidaceae.*
After Narcissus. Bulbous perennials.
Daffodil.
  asturiensis, a-stu-ree-*en*-sis. Of
  Asturia, Spain.
  bulbocodium, bul-bo-*ko*-dee-um.
  Woolly bulb. Hoop Petticoat
  Daffodil.
  conspicuus, kon-*spik*-ew-us.
  Conspicuous.
  cantabricus, kan-*tab*-ri-kus. Of
  Cantabria, Spain.
  cyclamineus, sik-la-*min*-ee-us.
  Cyclamen-like.
  jonquilla, jong-*kwil*-la. Slender
  leaves.
  minor, mi-nor. Smaller.

papyraceus, pa-pi-*ra*-see-us. Paper-
like.
poeticus, po-*e*-ti-kus. Of poets.
Poet's Narcissus, Pheasant-eye
Narcissus.
pseudonarcissus, soo-do-nar-*sis*-us.
False *Narcissus.* Wild Daffodil,
Trumpet Narcissus.
requienii, rek-wee-*en*-ee-ee. After
Requien. Rush-leaved Jonquil.
romieuxii, rom-*ew*-ee-ee. After
Romieux.
rupicola, roo-*pi*-ko-la. Growing on
rocks.
tazetta, ta-*ze*-ta. Small cup.
tenuifolius, ten-ew-i-*fo*-lee-us.
Slender-leaved.
italicus, i-*ta*-li-kus. Italian.
viridiflorus, vi-ri-di-*flo*-rus. With
green flowers.

Nasturtium, nas-*tur*-she-um.
Cruciferae. From L. *nasi tortium*

Narcissus pseudonarcissus

Nasturtium officinale

(twisted nose), after the smell of the leaves. Aquatic, perennial herb. Watercress.
officinale, o-fi-si-*na*-lee. Sold in shops. Watercress.

Nectaroscordum, nek-ta-ro-*skor*-dum. *Liliaceae.* From Gk. *nektar* (nectar) and *skordon* (garlic). Bulbous perennial.
siculum, sik-ew-lum. Of Sicily. Sicilian Honey Garlic.

Neillia, *neel*-ee-a. *Rosaceae.* After Dr. Patrick Neill. Deciduous shrubs.
sinensis, si-*nen*-sis. Of China.
thibetica, ti-*be*-ti-ka. Of Tibet.

Nelumbo, ne-*lum*-bo. *Nelumbonaceae.* The Sinhalese name. Deciduous, perennial, aquatic herbs. Lotus.
lutea, loo-tee-a. Yellow. American Lotus.
nucifera, new-*ki*-fe-ra. Nut-bearing. Sacred Lotus.

Nemesia, ne-*me*-see-a. *Scrophulariaceae.* From Gk. *nemesion.* Annual, perennial sub-shrubs.
floribunda, flo-ri-*bun*-da. Profusely flowering.
strumosa, stroo-*mo*-sa. With cush-ion-like swellings.
versicolor, ver-*si*-ko-lor. Variously coloured.

Nemophila, ne-*mo*-fi-la. *Hydrophyllaceae.* From Gk. *nemos* (glade) and *phileo* (love), growing in shady places. Annual herbs.
maculata, mak-ew-*la*-ta. Spotted (corolla). Five spot.
menziesii, men-*zeez*-ee-ee. After Menzies. Baby Blue Eyes.

Neoporteria, nee-o-por-*te*-ree-a. *Cactaceae.* After Carlos Porter.

chilensis, chil-*en*-sis. Of Chile.
subgibbosa, sub-ji-*bo*-sa. Somewhat swollen on one side.
villosa, vil-*lo*-sa. Softly hairy.

Neoregelia, nee-o-ree-*gel*-ee-a. *Bromeliaceae.* After Eduard Albert von Regel. Tender, evergreen perennial herbs.
carolinae, ka-ro-*leen*-ee. After Carolina. Blushing Bromeliad.
marmorata, mar-mo-*ra*-ta. Marbled (leaves). Marble Plant.
spectabilis, spek-*ta*-bi-lis. Spectacular. Painted Fingernail.

Nepenthes, nee-*pen*-theez. *Nepenthaceae.* From Gk. 'without care'. Tender, insectivorous, evergreen shrubs. Pitcher Plant.
gracilis, gra-si-lis. Graceful.
hookeriana, huk-a-ree-*a*-na. After Hooker.
maxima, max-i-ma. Largest.
rafflesiana, raf-lee-zee-*a*-na. After Sir Stamford Raffles.
ventricosa, ven-tri-*ko*-sa. Swollen on one side.

Nepeta, *ne*-pe-ta. *Labiatae.* L. name. Perennial herbs.
cataria, ka-*ta*-ree-a. Of cats. Catnip, Catmint.
x *faassenii,* far-*sen*-ee-ee. After J. H. Faassen.
grandiflora, grand-i-*flo*-ra. Large-flowered.
nervosa, ner-*vo*-sa. Conspicuously veined.

Nephrolepis, nef-ro-*lep*-is. *Oleandraceae.* From Gk. *nephros* (kid-ney) and *lepis* (scale). Tender, semi-evergreen ferns. Sword Fern.
cordifolia, kor-di-*fo*-lee-a. With heart-shaped leaves. Ladder Fern.

Nepeta cataria

exaltata, ex-al-*ta*-ta. Very tall. Boston Fern.
multiflora, mul-ti-*flo*-ra. Many-flowered. Asian Sword Fern.

Nerine, nee-*ree*-nee. *Amaryllidaceae.* After Nerine, a water nymph. Semi-hardy, bulbous herbs.
bowdenii, bow-*den*-ee-ee. After Mr Athelston Bowden.
filifolia, fil-i-*fo*-lee-a. With thread-like leaves.
flexuosa, flex-ew-*o*-sa. Wavy.
sarniensis, sar-nee-*en*-sis. Of Sarnia (Guernsey).
undulata, un-dew-*la*-ta. Wavy.

Nerium, *nee*-ree-um. *Apocynaceae.* Gk. name for Oleander. Tender, evergreen, poisonous shrub.
oleander, o-lee-*an*-der. From the Italian *oleandra.* Oleander.

Nertera, *ner*-te-ra. *Rubiaceae.* From Gk. *nerteros* (low down). Tender creeping herb.
depressa, dee-*pres*-sa. Flattened.
granadensis, gran-a-*den*-sis. Of Granada, Colombia. Bead Plant.

Nicandra, ni-*kan*-dra. *Solanaceae.* After Nikander of Colophon. Annual herb.
physalodes, fi-sal-*o*-deez. *Physalis*-like. Apple of Peru, Shoo Fly.

Nicotiana, nee-ko-tee-*a*-na. *Solanaceae.* After Jean Nicot. Annual and perennial herbs. Tobacco.
affinis, a-*fee*-nis. Related to.
alata, a-*la*-ta. Winged. Jasmine Tobacco.
glauca, glow-ka. Smooth. Tree Tobacco.
sylvestris, sil-*ves*-tris. Of woods.

Nidularium, need-ew-*la*-ree-um. *Bromeliaceae.* From L. *nidus* (nest). Tender, evergreen herbs.
carolinae, ka-ro-*leen*-ee. After Carolina.
fulgens, ful-jens. Shining (bracts).
innocentii, in-o-*sent*-ee-ee. After Pope Innocenti.
procerum, pro-*see*-rum. Tall.

Nierembergia, nee-e-ram-*berg*-ee-a. *Solanaceae.* After Juan Eusebio Nieremberg. Perennial herbs. Cupflower.
caerulea, see-*ru*-lee-a. Dark blue.
hippomanica, hip-o-*man*-i-ka. A plant horses eat.
violacea, vie-o-*la*-see-a. Violet.
repens, ree-pens. Creeping. Whitecup.

Nigella, ni-*jel*-la. *Ranunculaceae.* From L. *niger* (black), after the seeds. Annual herbs.
arvensis, ar-*ven*-sis. Of fields. Wild Fennel
damascena, dam-a-*see*-na. Of Damascus. Love-in-a-mist.

hispanica, his-*pa*-ni-ka. Of Spain.
Fennel Flower.
orientalis, o-ree-en-*ta*-lis. Eastern.
sativa, sa-*teev*-a. Cultivated. Black
Cumin.

Nolana, no-*la*-na. *Nolanaceae.* From
L. *nola* (little bell), after the corolla
shape. Perennial herbs.
acuminata, a-kew-mi-*na*-ta. Long-
pointed.
humifusa, hum-i-*few*-sa. Prostrate.
paradoxa, pa-ra-*dox*-a. Unusual.

Nomocharis, no-mo-*ka*-ris. *Liliaceae.*
From Gk. *nomos* (meadow) and *charis*
(grace). Bulbous perennials.
mairei, mair-ree-ee. After Maire.
pardanthina, par-dan-*theen*-a.
Pardanthina-like.
saluenensis, sal-ew-en-*en*-sis. From
near the Salween River, China.

Nopalxochia, no-pal-*ho*-kee-a.
Cactaceae. From a Mexican name.
ackermannii, a-ker-*man*-ee-ee. After
Georg Ackermann.
phyllanthoides, fil-lanth-*oi*-deez.
Phyllanthus-like.

Nothofagus, no-tho-*fay*-gus.
Fagaceae. From Gk. *notos* (southern)
and *fagus* (beech). Deciduous and
evergreen trees. Southern Beech.
antarctica, an-*tark*-ti-ka. Of
Antarctic regions.
betuloides, bet-ew-*loi*-deez. *Betula*-
like. Guindo Beech.
dombeyi, dom-bee-ee. After
Dombey.
fusca, fus-ka. Brown. Red Beech.
menziesii, men-*zeez*-ee-ee. After
Menzies. Silver Beech.
obliqua, o-*blee*-kwa. Oblique. Roble
Beech.
procera, pro-*see*-ra. Tall. Rauli Beech.

Notocactus, no-to-*kak*-tus. *Cactaceae.*
From Gk. *notos* (southern) and *Cactus*.
apricus, a-*pree*-kus. Sun-loving.
haselbergii, ha-sel-*berg*-ee-ee. After
Dr von Haselberg. Scarlet Ball
Cactus.
leninghausii, len-ing-*how*-zee-ee.
After Leninghaus. Golden Ball
Cactus.
mammulosus, mam-ew-*lo*-sus.
Bearing nipples.
ottonis, o-*to*-nis. After Friedrich
Otto.
scopa, sko-pa. Broom-like. Silver
Ball Cactus.

Notospartium, no-to-*spar*-tee-um.
Leguminosae. From Gk. *notos* (southern) and *Spartium.* Semi-hardy, leafless shrub.
carmichaeliae, kar-mie-*keel*-ee-ee.
Carmichaelia-like. Pink Broom.

Nuphar, *new*-far. *Nymphaeaceae.*
From the Arabic name. Deciduous,
perennial, aquatic herbs.
advena, ad-*ven*-a. Adventive.
Spatterdock.
lutea, loo-tee-a. Yellow. Yellow
Water Lily.

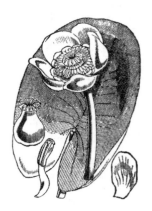

Nuphar lutea

Nymphaea alba

Nymphaea, nim-*fee*-a.
Nymphaeaceae. After Nymphe, a
water nymph. Deciduous, perennial,
aquatic herbs. Water Lily.
　alba, al-ba. White. European White
　Lily.
　caerulea, see-*ru*-lee-a. Dark blue.
　Blue Lotus.
　candida, kan-di-da. White.
　capensis, ka-*pen*-sis. Of the Cape of
　Good Hope. Cape Blue Water Lily.

x *marliacea,* mar-lee-*a*-see-a. After
Joseph Latour Marliac.
odorata, o-do-*ra*-ta. Scented.
Fragrant Water Lily.
pygmaea, pig-*mee*-a. Dwarf.
stellata, ste-*la*-ta. Star-like.
tetragona, tet-ra-*gon*-a. Four-angled.
Pygmy Water Lily.

Nymphoides, nimf-*oi*-deez.
Menyanthaceae. Nymphaea-like.
Deciduous, perennial, aquatic herbs.
　aquatica, a-*kwa*-ti-ka. Growing in
　water. Fairy Water Lily.
　cordata, kor-*da*-ta. Heart-shaped.
　indica, in-di-ka. Of India. Water
　Snowflake.
　peltata, pel-*ta*-ta. Shield-shaped.
　Yellow Floating Heart.

Nyssa, *nie*-sa. *Nyssaceae.* After Nyssa,
a water nymph. Deciduous trees.
　sinensis, si-*nen*-sis. Of China.
　sylvatica, sil-*va*-ti-ka. Of woods.
　Black Gum, Tupelo.

O

Ochna, *ok*-na. *Ochnaceae*. From Gk. *ochne* (Wild Pear). Tender, evergreen trees and shrubs. Bird's Eye bush.
    serrulata, se-ru-*la*-ta. With small teeth

Ocimum, *o*-si-mum. *Labiatae*. From Gk. *okimon* (aromatic herb). Annual herbs.
    basilicum, ba-*si*-li-kum. Royal. Basil.
    tenuiflorum, ten-ew-ee-*flo*-rum. With slender flowers. Holy Basil.

Odontoglossum, o-don-to-*glos*-um. *Orchidaceae*. From Gk. *odontos* (tooth) and *glossa* (tongue), after the toothed lip. Epiphytic or lithophytic greenhouse orchids.
    cervantesii, ser-van-*tez*-ee-ee. After Professor Cervantes.
    cordatum, kor-*da*-tum. Heart-shaped (lip).
    crispum, kris-pum. Wavy-edged. Lace Orchid.
    grande, gran-dee. Large. Tiger Orchid.
    harryanum, ha-ree-*a*-num. After Sir Harry Veitch.
    pulchellum, pul-*kel*-um. Pretty.
    rossii, ros-ee-ee. After John Ross.
    triumphans, tri-*um*-fanz. Splendid.

Oenothera, ee-no-*the*-ra. *Onagraceae*. From Gk. *oinos* (wine) and *thera* (imbibing). Annual, biennial and perennial herbs.
    acaulis, a-*kaw*-lis. Stemless.
    berlandieri, ber-lan-dee-*e*-ree. After J. L. Berlandier.
    biennis, bee-*en*-is. Biennial.

Common Evening Primrose.
    caespitosa, see-spi-*to*-sa. Tufted.
    grandiflora, grand-i-*flo*-ra. Large-flowered.
    laciniata, la-sin-ee-*a*-ta. Deeply cut (leaves).
    macrocarpa, mak-ro-*kar*-pa. Large-fruited. Ozark Sundrops.
    missouriensis, mi-sur-ree-*en*-sis. Of Missouri.
    perennis, pe-*ren*-is. Perennial. Sundrops.
    speciosa, spes-ee-*o*-sa. Showy. White Evening Primrose.
    riparia, ree-*pa*-ree-a. Of river banks.

Oenothera biennis

Olea, *o*-lee-a. *Oleaceae*. L. for olive. Semi-hardy, evergreen tree.
    europaea, ew-ro-*pee*-a. European. Edible Olive.

Olearia, o-lee-*a*-ree-a. *Compositae*. Origin unknown. Evergreen trees and shrubs. Daisy Bush.

albida, al-bi-da. Whitish.

avicenniifolia, a-vi-sen-ee-i-*fo*-lee-a. With *Avicennia*-like leaves.

chathamica, cha-*tam*-i-ka. Of the Chatham Islands.

frostii, frost-ee-ee. After Charles Frost.

x *haastii, harst*-ee-ee. After Sir Johann von Haast.

ilicifolia, ee-li-si-*fo*-lee-a. *Ilex*-leaved.

macrodonta, mak-ro-*don*-ta. With large teeth (leaves).

x *mollis, mol*-lis. Softly hairy. New Zealand Holly.

nummulariifolia, num-ew-la-ree-i-*fo*-lee-a. With coin-shaped leaves.

phlogopappa, flog-o-*pa*-pa. With a *Phlox*-like pappus.

splendens, *splen*-denz. Splendid.

x *scilloniensis,* si-lon-ee-*en*-sis. Of the Scilly Isles.

traversii, tra-*verz*-ee-ee. After W. T. L. Travers.

virgata, vir-*ga*-ta. Twiggy.

Omphalodes, om-fa-*lo*-deez. *Boraginaceae.* From Gk. *omphalos* (navel), after the seed shape. Annual and perennial herbs. Navelwort, Navelseed.

    cappadocica, kap-a-*do*-si-ka. Of Cappadocia,Turkey.

    japonica, ja-*pon*-i-ka. Of Japan.

    linifolia, lin-i-*fo*-lee-a. *Linum*-leaved. Venus's Navelwort.

    luciliae, loo-*sil*-ee-ee. After Lucile Boissier.

    verna, ver-na. Of Spring. Creeping Forget-me-not.

Oncidium, on-*sid*-ee-um. *Orchidaceae.* From Gk. *onkos* (tumour), after the swelling on the lip. Epiphytic, lithophytic and terrestrial greenhouse orchids.

abortivum, ab-or-*te*-vum. Imperfect.

altissimum, al-*tis*-i-mum. Tallest.

aureum, aw-ree-um. Golden.

cheirophorum, ky-*ro*-fo-rum. Hand-bearing. Colombia Buttercup.

crispum, kris-pum. Finely wavy.

cucullatum, kuk-ew-*la*-tum. Hood-like.

flexuosum, flex-ew-*o*-sum. Tortuous.

incurvum, in-*kur*-vum. Incurved.

longifolium, long-i-*fo*-lee-um. Long-leaved.

luridum, loo-ri-dum. Pale yellow.

macranthum, ma-*kranth*-um. Large-flowered.

marshallianum, mar-shal-ee-*a*-num. After Mr W Marshall.

ornithorhyncum, or-ni-thor-*in*-kum. Like a bird's beak.

papilio, pa-*pil*-ee-o. Butterfly. Butterfly Orchid.

pulchellum, pul-*kel*-um. Pretty.

pumilum, pew-mi-lum. Dwarf.

sarcodes, sar-*ko*-deez. Flesh-like.

splendidum, splen-di-dum. Splendid.

tigrinum, ti-*gree*-num. Striped like a tiger.

Onoclea, o-*nok*-lee-a. *Athyriaceae.* From Gk. *onos* (vessel) and *kleio* (close). Deciduous Fern.

    sensibilis, sen-*si*-bi-lis. Sensitive. Sensitive Fern. Bead Fern.

Ononis, o-*no*-nis. *Leguminosae.* Gk. name. Perennial herbs and deciduous or semi-evergreen sub-shrubs.

    aragonensis, a-ra-gon-*en*-sis. Of Aragon.

    fruticosa, froo-ti-*ko*-sa. Shrubby.

    repens, ree-penz. Creeping.

    rotundifolia, ro-tun-di-*fo*-lee-a. With rounded leaves.

Onopordum, o-no-*por*-dum. *Compositae.* From Gk. *onos* (ass) and *perdo* (consume). Annual, biennial and perennial herbs.

acanthium, a-*kanth*-ee-um. Spiny. Giant Thistle.

illyricum, i-*li*-ri-kum. Of Illyria.

nervosum, ner-*vo*-sum. Veined.

Onosma, o-*nos*-ma. *Boraginaceae.* From Gk. *onos* (ass) and *osme* (smell). Annual and perennial herbs and sub-shrubs.

alborosea, al-bo-*ros*-ee-a. White and rose-coloured.

echioides, e-kee-*oi*-deez. *Echium*-like.

stellulata, stel-ew-*la*-ta. With small stars.

taurica, *taw*-ri-ka. Of the Crimea. Golden Drop.

Ophiopogon, o-fee-o-*po*-gon. *Liliaceae.* From Gk. *ophis* (snake) and *pogon* (beard). Evergreen, perennial herbs with grass-like foliage.

japonicus, ja-*pon*-i-kus. Of Japan.

planiscapus, plan-i-*ska*-pus. With a flat scape.

Ophrys, *of*-ris. *Orchidaceae.* Gk. name. Hardy, terrestrial, tuberous orchids.

apifera, a-*pi*-fe-ra. Bee-bearing (the labellum resembles a bee). Bee Orchid.

fusca, *fus*-ka. Brown.

lutea, *loo*-tee-a. Yellow. Yellow Tree Orchid.

speculum, *spek*-ew-lum. A mirror.

Oplismenus, op-*lis*-men-us. *Gramineae.* From Gk. *hoplismos* (weapon). Tender trailing grass.

compositus, kom-*po*-si-tus. Compound.

hirtellus, hir-*tel*-us. Rather hairy. Basket Grass.

Ophrys apifera

Opuntia, o-*pun*-tee-a. *Cactaceae.* From Gk. Prickly Pear.

azurea, a-*zew*-ree-a. Sky-blue.

basilaris, ba-si-*la*-ris. Basal.

bigelovii, big-a-*lov*-ee-ee. After Jacob Bigelow. Teddy Bear Cholla.

cylindrica, si-*lin*-dri-ka. Cylindrical.

decumbens, dee-*kum*-benz. Prostrate.

ficus-indica, *fi*-kus-*in*-di-ka. Fig of India.

humifusa, hum-i-*few*-sa. Low-growing.

humilis, *hum*-i-lis. Low-growing.

imbricata, im-bri-*ka*-ta. Densely overlapping.

leucotricha, loo-*ko*-tri-ka. With white hairs. Mexico.

microdasys, mik-ro-*das*-is. Small and shaggy.

rufida, *roo*-fi-da. Reddish.

ovata, o-*va*-ta. Ovate.

polyacantha, po-lee-a-*kanth*-a. Many-spined.

robusta, ro-*bus*-ta. Robust.

subulata, soo-bew-*la*-ta. Awl-shaped (leaves).

sulphurea, sul-*fu*-ree-a. Sulphur-yel-

low (flowers).
tunicata, tun-i-*ka*-ta. Coated.
verschaffeltii, vair-sha-*felt*-ee-ee.
After Verschaffelt.
vulgaris, vul-*ga*-ris. Common.

Orchis mascula

Orchis, *or*-kis. *Orchidaceae.* Gk.
name. Hardy, terrestrial orchids.
  elata, e-*la*-ta. Tall. Robust Marsh
  Orchid.
  foliosa, fo-lee-*o*-sa. Leafy.
  Madeiran Orchid.
  fuchsii, few-shee-ee. After Fuchs.
  Common Spotted Orchid.
  incarnata, in-kar-*na*-ta. Pink. Early
  Marsh Orchid.
  mascula, mas-kew-la. Male. Early
  Purple Orchid.
  militaris, mee-li-*ta*-ris. Like a soldier. Military Orchid.
  purpurea, pur-*pur*-ree-a. Purple.
  Lady Orchid.
  spectabilis, spek-*ta*-bi-lis.
  Spectacular.

Origanum, o-ree-*ga*-num. *Labiatae.*
From Gk. *oros* (mountain) and *ganos*

(beauty). Deciduous sub-shrubs and
perennial herbs. Marjoram, Oregano.
  amanum, a-*ma*-num. Of the Amanus
  Mountains.
  dictamnus, dik-*tam*-nus. Gk. name.
  Mountains. Hop Marjoram.
  laevigatum, lee-vi-*ga*-tum. Smooth.
  libanoticum, li-ba-*no*-ti-kum. Of
  Lebanon.
  marjorana, mar-jo-*ra*-na. From old
  Gk. *amarakus.* Sweet Marjoram.
  onites, o-*nie*-teez. Gk. name for a
  kind of marjoram. Pot Marjoram.
  rotundifolium, ro-tun-di-*fo*-lee-um.
  Round-leaved.
  vulgare, vul-*ga*-ree. Common. Wild
  Marjoram, Pot Marjoram, Oregano.

Ornithogalum, or-ni-*tho*-ga-lum.
Liliaceae. From Gk. *ornis* (bird) and
gala (milk). Bulbous, perennial herbs.
  arabicum, a-*ra*-bi-kum. Of Arabia.
  balansae, ba-*lan*-zee. After Benedict
  Balansa.
  montanum, mon-*ta*-num. Of mountains.

Ornithogalum nutans

narbonense, nar-bon-*en*-see. Of
Narbonne.
nutans, new-tanz. Nodding (flowers).
oligophyllum, ol-i-go-*fil*-um. With
few leaves.
thyrsoides, thur-*soi*-deez. With flowers in a thyrse.
umbellatum, um-bel-*a*-tum.
Umbelled. Star of Bethlehem.

Orontium, o-*ron*-tee-um. *Araceae.*Of
the region of the River Orontes.
Deciduous, perennial, aquatic herb.
aquaticum, a-*kwa*-ti-kum. Growing
in water. Golden Club.

Osmanthus, os-*manth*-us. *Oleaceae.*
From Gk. *osme* (fragrance) and *anthos*
(flower). Evergreen trees and shrubs.
americanus, a-me-ri-*ka*-nus. Of
America. Devilwood.
armatus, ar-*ma*-tus. Spiny (leaves).
x *burkwoodii,* burk-*wud*-ee-ee. After
Burkwood.
decorus, de-*ko*-rus. Beautiful.
delavayi, del-a-*vay*-ee. After
Delavay.
fragrans, fra-granz. Fragrant.
Fragrant Olive, Sweet Tea.
heterophyllus, he-te-ro-*fil*-lus. With
variable leaves. Holly Olive,
Chinese Holly.
yunnanensis, yoo-nan-*en*-sis. Of
Yunnan.

Osmaronia, os-ma-*ro*-nee-a.
Rosaceae. From Gk. *osme* (fragrance)
and *Aronia* (Chokeberry). Deciduous
shrub.
cerasiformis, se-ra-si-*form*-is.
Cherry-shaped (fruit). Oso Berry,
Oregon Plum.

Osmunda, os-*mun*-da. *Osmundaceae.*
Origin unknown. Deciduous ferns.

cinnamomea, sin-a-*mo*-mee-a.
Cinnamon-coloured. Cinnamon Fern,
Buckhorn.
claytonia, klay-*ton*-ee-a. After John
Clayton. Interrupted Fern.
regalis, ree-*ga*-lis. Regal. Royal
Fern, Flowering Fern.

Osteospermum, ost-ee-o-*sperm*-um.
Compositae. From Gk. *osteon* (bone)
and *sperma* (seed). Semi-hardy subshrubs and perennial herbs.
barberiae, bar-*be*-ree-ee. After Mrs
Barber.
ecklonis, ek-*lon*-is. After Christian
Ecklon.
jucundum, joo-*kun*-dum. Pleasing.

Ostrowskia, os-*trov*-skee-a.
Campanulaceae. After Michael
Nicholazewitsch von Ostrowsky.
Perennial herb.
magnifica, mag-*ni*-fi-ka.
Magnificent. Giant Bellflower.

Osmunda regalis

Ostrya, *os*-tree-a. *Betulaceae*. From Gk. *ostrys* (scale). Deciduous trees.
   carpinifolia, kar-pin-i-*fo*-lee-a. *Carpinus*-like leaves. Hop Hornbeam.
   japonica, ja-*pon*-i-ka. Of Japan.
   virginiana, vir-jin-ee-*a*-na. Of Virginia. Iron Wood, Eastern Hop Hornbeam.

Ourisia, ow-*ris*-ee-a. *Scrophulariaceae*. After Governor Ouris of the Falkland Islands. Evergreen perennial herbs and sub-shrubs.
   caespitosa, see-spi-*to*-sa. Tufted.
   coccinea, kok-*kin*-ee-a. Scarlet.
   elegans, *e*-le-ganz. Elegant.
   macrocarpa, mak-ro-*kar*-pa. Large-fruited.
   macrophylla, mak-ro-*fil*-la. Large-leaved.

Oxalis, ox-*a*-lis. *Oxalidaceae*. From Gk. *oxys* (acid). Hardy and tender herbs. Sorrel, Shamrock.
   acetosella, a-see-to-*se*-la. With acid leaves. Wood Sorrel.
   adenophylla, a-den-o-*fil*-la. With glandular leaves.
   articulata, ar-tik-ew-*la*-ta. Jointed.
   chrysantha, kris-*anth*-a. With golden flowers.
   depressa, dee-*pres*-a. Flattened.
   enneaphylla, en-ee-a-*fil*-la. With nine leaflets. Scurvy Grass.
   hirta, *hir*-ta. Hairy.
   laciniata, la-sin-ee-*a*-ta. Deeply cut.
   latifolia, la-ti-*fo*-lee-a. Broad-leaved.
   lobata, lo-*ba*-ta. Lobed.
   magellanica, ma-ge-*lan*-i-ka. From the region of the Magellan Straits.
   oregona, o-ree-*go*-na. Of Oregon. Redwood Sorrel.
   ortgiesii, ort-*geez*-ee-ee. After Eduard Ortgies. Tree Oxalis.

Oxalis acetosella

   pes-caprae, pes-*kap*-ree. Like a goat's foot. Bermuda Buttercup.
   tetraphylla, tet-ra-*fil*-a. With four leaves. Lucky Clover.
   violacea, vie-o-*la*-see-a. Violet. Violet Wood Sorrel.

Oxycoccus, ox-i-*kok*-us. *Ericaceae*. From Gk. *oxys* (acid) and *kokkos* (round berry). Prostrate, evergreen shrubs. Cranberry.
   macrocarpus, mak-ro-*kar*-pus. With large fruit. American Cranberry.
   palustris, pa-*lus*-tris. Growing in marshes. European Cranberry.

Oxydendrum, ox-i-*den*-drum. *Ericaceae*. From Gk. *oxys* (acid) and *dendron* (tree). Deciduous tree.
   arboreum, ar-*bor*-ee-um. Tree-like. Sorrel Tree, Sourwood.

Oxypetalum, ox-i-*pe*-ta-lum. *Asclepiadaceae*. From Gk. *oxys* (sharp) and *petalum* (petal). Tender climber.
   caeruleum, see-*ru*-lee-um. Dark blue.

Ozothamnus, o-zo-*tham*-nus.
Compositae. From Gk. *ozo* (smell) and
thamnos (shrub). Evergreen shrubs.
  depressus, dee-*pres*-sus. Flattened.
  ledifolius, le-di-*fo*-lee-us. With
Ledum-like leaves.
  purpurascens, pur-pur-*ras*-enz.
Purplish.
  rosmarinifolius, ros-ma-reen-i-*fo*-
lee-us. With *Rosmarinus*-like leaves.

P

Pachyphragma, pa-kee-*frag*-ma.
Cruciferae. From Gk. *pachys* (thick)
and *phragma* (screen), after the ribbed
seed pods. Perennial herb.
  macrophyllum, mak-ro-*fil*-lum.
  Large-leaved.

Pachyphytum, pa-kee-*fi*-tum.
Crassulaceae. From Gk. *pachys*
(thick) and *phyton* (plant).
Tender, perennial succulents.
  compactum, com-*pak*-tum.
  Compact.
  oviferum, o-*vi*-fe-rum. Egg-bearing
  (leaf shape). Moonstones.
  viride, vi-ri-dee. Green.

Pachysandra, pa-kis-*an*-dra.
Buxaceae. From Gk. *pachys* (thick)
and *andros* (male), after the thick stamens. Evergreen creeping perennials
and sub-shrubs.
  axillaris, ax-il-*la*-ris. In the leaf axil.
  procumbens, pro-*kum*-benz.
  Prostrate.
  terminalis, ter-mi-*na*-lis. Terminal
  (flower spikes).

Pachystachys, pa-kee-*sta*-kis.
Acanthaceae. From Gk. *pachys* (thick)
and *stachys* (spike). Tender evergreen
perennials and shrubs.
  coccinea, kok-*kin*-ee-a. Scarlet.
  Cardinal's Guard.
  lutea, loo-tee-a. Yellow (bracts).

Paeonia, pee-*on*-ee-a. *Paeoniaceae.*
After Gk. Paion, physician to the gods.
Herbaceous perennials and deciduous
shrubs. Peony.
  anomala, a-*nom*-a-la. Unusual.

Paeonia officinalis

  arietina, a-ree-e-*tee*-na. Like a ram's
  head.
  clusii, clooz-ee-ee. After Clusius.
  delavayi, de-la-*vay*-ee. After
  Delavay.
  lactiflora, lak-ti-*flo*-ra. With milky
  flowers.
  lobata, lo-*ba*-ta. Lobed.
  lutea, loo-tea. Yellow.
  mascula, mas-kew-la. Male.
  mlokosewitschii, mlo-ko-se-*vich*-ee-
  ee. After Ludwig Franzevich
  Mlokosewitsch.
  obovata, ob-*ova*-ta. Obovate.
  officinalis, o-fi-si-*na*-lis. Sold in shops.
  peregrina, pe-re-*green*-a. Foreign.
  potaninii, po-ta-*nin*-ee-ee. After
  Nicolaevich Potanin.
  suffruticosa, suf-froo-ti-*ko*-sa. Sub-
  shrubby.
  tenuifolia, ten-ew-i-*fo*-lee-a. With
  slender leaves.
  veitchii, veech-ee-ee. After the
  Veitch nursery.
  wittmanniana, vit-man-ee-*a*-na.
  After Wittmann.

Paliurus, pa-li-*ew*-rus. *Rhamnaceae.*
Gk. name. Deciduous shrub.
spina-christi, speen-a-*kris*-tee.
Christ's Thorn.

Pancratium, pan-*krat*-ee-um.
Amaryllidaceae. From Gk. *pan* (all)
and *kratos* (potent). Bulbous perennials.
canariense, ka-na-ree-*en*-see. Of the
Canary Islands.
illyricum, i-*li*-ri-kum. Of Illyria.
maritimum, ma-*ri*-ti-mum. Growing
near the sea.

Pandanus, *pan*-da-nus. *Pandanaceae.*
From Malayan *pandan.* Tender, evergreen shrubs and trees. Screw Pine.
odoratissimus, o-do-ra-*tis*-i-mus.
Highly scented.
pygmaeus, pig-*mee*-us. Dwarf.
sanderi, san-da-ree. After the Sander
nursery.
tectorius, tek-*taw*-re-us. On roofs.
veitchii, veech-ee-ee. After the
Veitch nursery.

Pandorea, pan-*do*-ree-a.
Bignoniaceae. After the mythological
Pandora. Tender, evergreen climbers.
jasminoides, jas-min-*oi*-deez.
Jasmine-like. Bower Plant.
pandorana, pan-do-*ra*-na. After
Pandora. Wonga-wonga Vine.

Panicum, *pa*-ni-kum. *Gramineae.* L.
for millet. Annual or perennial grasses.
capillare, ka-pi-*la*-ree. Hair-like. Old
Witch Grass.
miliaceum, mi-lee-*a*-see-um. Millet-
like. Millet, Hog Millet.
virgatum, vir-*ga*-tum. Wand-like.
Switch Grass

Papaver, pa-*pa*-ver. *Papaveraceae.* L.
for poppy. Annual, biennial and peren-

nial herbs. Poppy.
alpinum, al-*pie*-num. Alpine.
atlanticum, at-*lan*-ti-kum. Of the
Atlantic coast.
commutatum, kom-ew-*ta*-tum.
Changeable.
glaucum, glow-kum. Glaucous
(leaves). Tulip Poppy.
nudicaule, new-di-*kaw*-lee. Bare-
stemmed. Arctic Poppy, Icelandic
Poppy.
orientale, o-ree-en-*ta*-lee. Eastern.
Oriental Poppy.
pilosum, pi-*lo*-sum. Hairy.
rhoeas, ree-as. Corn Poppy, Flanders
Poppy.
rupifragum, roo-*pi*-fra-gum.
Growing in rocks. Spanish Poppy.
somniferum, som-*ni*-fe-rum. Sleep-
bearing. Opium Poppy.
spicatum, spee-*ka*-tum. With flowers
in spikes.

Papaver somniferum

Paphiopedilum, pa-fee-o-*pe*-di-lum.
Orchidaceae. From Gk. Paphos (temple where Aphrodite was worshipped)
and *pedilon* (slipper). Greenhouse

orchids. Venus' Slipper Orchid.
bellatulum, be-*la*-tew-lum. Pretty.
callosum, ka-*lo*-sum. Calloused
(petals).
fairrieanum, fair-ree-*a*-num. After
Mr Fairrie.
niveum, *niv*-ee-um. Snow-white
(lip).
purpuratum, pur-pur-*ra*-tum.
Purplish.
sukhakulii, soo-ka-*koo*-lee-ee. After
P. Sukhakuli.
venustum, ve-*nus*-tum. Handsome.

Paradisea, pa-ra-*diz*-ee-a. *Liliaceae*.
After Count Giovanni Paradisi.
Perennial herb.
liliastrum, lil-ee-*as*-trum. *Lilium*-
like. St Bruno's Lily, Paradise Lily.
lusitanicum, loo-sit-*an*-i-kum. Of
Portugal.

Parahebe, pa-ra-*hee*-bee.
Scrophulariaceae. From Gk. *para*
(close to) and *Hebe*. Dwarf, evergreen
or semi-evergreen sub-shrubs and
shrubs.
catarractae, ka-ta-*rak*-tee. Of water-
falls.
decora, de-*ko*-ra. Beautiful.
lyallii, lie-*al*-ee-ee. After David
Lyall.
perfoliata, per-fo-li-*a*-ta. With the
leaf surrounding the stem. Digger's
Speedwell.

Parkinsonia, par-kin-*son*-ee-a.
Leguminosae. After John Parkinson.
Tender, evergreen trees and shrubs.
aculeata, a-kew-lee-*a*-ta. Prickly.
Jerusalem Thorn.
florida, *flo*-ri-da. Flowering.

Parnassia, par-*nas*-ee-a.
Saxifragaceae. After Mt Parnassus,
Greece. Perennial herb. Grass of

Parnassia palustris

Parnassus, Bog Star.
palustris, pa-*lus*-tris. Of marshes.

Parochetus, pa-*ro*-ke-tus.
Leguminosae. From Gk. *para* (near)
and *ochetus* (brook). Semi-hardy, evergreen perennial herb.
communis, kom-*ew*-nis. Common.
Shamrock Pea, Blue Oxalis.

Parodia, pa-*ro*-dee-a. *Cactaceae*.
After Lorenzi Parodi.
chrysacanthion, kris-a-*kanth*-ee-on.
With golden spines.
nivosa, ni-*vo*-sa. Snow-white
(spines).
sanguiniflora, sang-gwin-i-*flo*-ra.
With blood-red flowers.

Parrotia, pa-*rot*-ee-a.
Hamamelidaceae. After F. W. Parrot.
Deciduous tree.
persica, *per*-si-ka. Of Persia.
Ironwood, Irontree.

Parrotiopsis, pa-rot-ee-*op*-sis.
Hamamelidaceae. From *Parrotia* and
Gk. -*opsis* (resemblance). Deciduous
tree or shrub.

jacquemontiana, zhak-a-mont-ee-*a*-na. After Victor Jacquemont.

Parthenocissus, par-then-o-*sis*-us. *Vitaceae.* From Gk. *parthenos* (virgin) and *kissos* (ivy). Deciduous climbers. Virginia Creeper.

henryana, hen-ree-*a*-na. After Augustine Henry.

quinquefolia, kwing-kwee-*fo*-lee-a. With five leaves. Virginia Creeper, Woodbine.

thomsonii, tom-*son*-ee-ee. After Thomson.

tricuspidata, tri-kus-pi-*da*-ta. Three-pointed (leaves). Japanese Creeper, Boston Ivy.

Passiflora, pa-si-*flo*-ra. *Passifloraceae.* From L. *passio* (passion) and *flos* (flower). Half-hardy to tender, evergreen climbers. Passion Flower.

x *allardii,* a-*lard*-ee-ee. After Edgar Allard.

antioquiensis, an-tee-o-kwee-*en*-sis. Of Antioquia, Colombia. Banana Passion Fruit.

caerulea, see-*ru*-lee-a. Dark blue. Passion Flower, Blue Passion Flower.

coccinea, kok-*kin*-ee-a. Scarlet. Red Granadilla, Red Passion Flower.

edulis, e-*dew*-lis. Edible (fruit). Granadilla, Passion Fruit.

x *exoniensis,* ex-o-nee-*en*-sis. Of Exeter.

laurifolia, law-ri-*fo*-lee-a. *Laurus*-leaved. Yellow Granadilla. Jamaican Honeysuckle.

manicata, man-i-*ka*-ta. Long-sleeved. Red Passion Flower.

mixta, mix-ta. Mixed.

mollissima, mol-*lis*-i-ma. Very softly hairy.

quadrangularis, kwod-rang-gew-*la*-ris. Four-angled (shoots). Giant Granadilla.

racemosa, ra-see-*mo*-sa. With flowers in racemes.

sanguinea, sang-*gwin*-ee-a. Red. Red Passion Flower.

vitifolia, vee-ti-*fo*-lee-a. *Vitis*-leaved.

Paulownia, paw-*lo*-nee-a. *Scrophulariaceae.* After Anna Paulowna, daughter of Czar Paul I. Deciduous trees.

fortunei, for-*tewn*-ee-ee. After Robert Fortune.

imperialis, im-peer-ee-*a*-lis. Showy.

lilacina, li-la-*seen*-a. Lilac (flowers).

tomentosa, to-men-*to*-sa. Hairy (leaves).

Pavonia, pa-*von*-ee-a. *Malvaceae.* After Jose Pavon. Tender, evergreen shrub.

hastata, has-*ta*-ta. Spear-shaped.

multiflora, mul-ti-*flo*-ra. Many-flowered.

Paxistima, pax-*i*-sti-ma. *Celastraceae.* From Gk. *pachys* (thick) and *stigma.* Evergreen shrubs.

canbyi, kan-bee-ee. After William Canby. Cliff Green.

myrtifolia, mur-ti-*fo*-lee-a. *Myrtus*-leaved. Oregon Boxwood.

Pedilanthus, pe-di-*lanth*-us. *Euphorbiaceae.* From Gk. *pedilon* (slipper) and *anthos* (flower). Tender perennial succulents.

tithymaloides, ti-thee-ma-*loi*-deez. *Tithymalus*-like.

Pelargonium, pel-ar-*gon*-ee-um. *Geraniaceae.* From Gk. *pelargos* (stork), after the beak of the fruit. Tender, perennial, mostly evergreen herbs and shrubs. Geranium.

crispum, kris-pum. Finely wavy (leaves). Lemon Geranium.
denticulatum, den-tik-ew-*la*-tum. Toothed (leaves). Fern-leaf Geranium.
x *domesticum,* do-*mes*-ti-kum. Cultivated. Regal Pelargonium.
x *fragrans, fra*-granz. Fragrant (leaves). Nutmeg Pelargonium.
fulgidum, ful-ji-dum. Shining.
graveolens, gra-*vee*-o-lenz. Aromatic. Rose Geranium.
x *hortorum,* hor-*to*-rum. Of gardens. Geranium, Zonal Pelargonium.
odoratissimum, o-do-ra-*tis*-i-mum. Highly scented (leaves). Apple Geranium.
peltatum, pel-*ta*-tum. Shield-shaped (leaves). Ivy Geranium.
quercifolium, kwer-ki-*fo*-lee-um. *Quercus*-leaved. Oak-leaved Geranium.
tetragonum, tet-ra-*go*-num. Four-angled (stems).
tomentosum, to-men-*to*-sum. Hairy.
triste, tris-tee. Sad.

Pellaea, pe-*lee*-a. *Adiantaceae*. From Gk. *pellaios* (dark), after the dark stalks. Ferns.
atropurpurea, at-ro-pur-*pur*-ree-a. Deep purple. Purple Cliff Brake.
rotundifolia, ro-tund-i-*fo*-lee-a. With round leaves. Button Fern.

Pellionia, pe-li-*on*-ee-a. *Urticaceae*. After Alphonse Pellion. Tender, evergreen, creeping perennial herbs.
daveauana, da-vo-*a*-na. After Jules Daveau. Trailing Watermelon Begonia.
pulchra, pul-kra. Pretty. Rainbow Vine.
repens, ree-penz. Creeping.

Peltiphyllum, pel-ti-*fil*-lum. *Saxifragaceae*. From Gk. *pelte* (shield) and *phyllon* (leaf), after the shield-like leaves. Perennial herb.
peltatum, pel-*ta*-tum. Shield-shaped (leaves). Umbrella Plant.

Pennisetum, pen-i-*se*-tum. *Gramineae*. From L. *penna* (feather) and *seta* (bristle). Perennial grasses.
alopecuroides, a-lo-pek-ew-*roi*-deez. *Alopecurus*-like. Chinese Pennisetum.
compressum, kom-*pres*-um. Compressed.
setaceum, se-*ta*-see-um. Bristly. Fountain Grass.
villosum, vi-*lo*-sum. Softly hairy. Feathertop.

Penstemon, pen-*ste*-mon. *Scrophulariaceae*. From Gk. *pente* (five) and *stemon* (stamen). Perennial herbs, and sub-shrubs.
barbatus, bar-*ba*-tus. Bearded.
davidsonii, day-vid-*son*-ee-ee. After Davidson.
fruticosus, froo-ti-*ko*-sus. Shrubby.
hartwegii, hart-*weg*-ee-ee. After Carl Theodore Hartweg.
heterophyllos, he-te-ro-*fil*-lus. With variable leaves. Foothill Penstemon.
newberryi, new-*be*-ree-ee. After J. S. Newberry. Mountain Pride.
humilior, hu-*mil*-ee-or. Low-growing.
nitidus, ni-ti-dus. Shining
ovatus, o-*va*-tus. Ovate (leaves). Broad-leaved Penstemon.
pinifolius, pie-ni-*fo*-lee-us. With *Pinus*-like leaves.
rupicola, roo-*pi*-ko-la. Growing on rocks.
virens, vi-renz. Green.

Pentas, *pen*-tas. *Rubiaceae*. From Gk. *pentas* (group of five). Tender, evergreen perennials and shrubs.

carnea, kar-nee-a. Flesh-coloured.
lanceolata, lan-see-o-*la*-ta. Lance-
shaped (leaves). Star Cluster.

Peperomia, pe-pe-*rom*-ee-a.
Piperaceae. From Gk. *peperi* (pepper)
and *homoios* (resembling). Tender,
evergreen, perennial herbs. Radiator
Plant.
   argyreia, ar-ji-*ree*-a. Silvery
(leaves). Watermelon Pepper.
   caperata, ka-pe-*ra*-ta. Wrinkled
(leaves).
   glabella, gla-*bel*-la. Rather smooth.
   magnoliifolia, mag-nol-ee-i-*fo*-lee-a.
Magnolia-leaved.
   obtusifolia, ob-tew-si-*fo*-lee-a.
Blunt-leaved. Baby Rubber Plant.
   scandens, skan-denz. Climbing.

Pereskia, pe-*res*-kee-a, *Cactaceae.*
After Nicholas de Peiresc. Deciduous
Cacti.
   aculeata, a-kew-lee-*a*-ta. Prickly.
   grandifolia, grand-i-*fo*-lee-a. Large-
leaved.

Perilla, pe-*ril*-la. *Labiatae.* Origin
unknown. Annual, half-hardy herb.
   frutescens, froo-*tes*-enz. Shrubby.

Peristrophe, pe-*ri*-stro-fee.
Acanthaceae. From Gk. *peri* (around)
and *strophe* (turning). Tender, perennial sub-shrubs.
   angustifolia, an-gust-i-*fo*-lee-a.
Narrow-leaved.
   hyssopifolia, hi-so-pi-*fo*-lee-a.
Hyssopus-leaved.
   speciosa, spes-ee-*o*-sa. Showy.

Pernettya, per-*net*-ee-a. *Ericaceae.*
After Antoine Joseph Pernetty.
Evergreen shrubs.
   mucronata, mew-kron-*a*-ta. Pointed
(leaves).

prostrata, pros-*tra*-ta. Prostrate.
pumila, pew-mi-la. Dwarf.

Perovskia, pe-*rof*-skee-a. *Labiatae.*
After V. A. Perovsky. Deciduous sub-
shrubs.
   atriplicifolia, a-tri-pli-ki-*fo*-lee-a.
Atriplex-leaved.

Petasites, pe-ta-*si*-teez. *Compositae.*
From Gk. *petasos* (hat). Perennial
herbs.
   fragrans, fra-granz. Fragrant. Winter
Heliotrope.
   japonicus, ja-*pon*-i-kus. Of Japan.

Petrea, *pet*-ree-a. *Verbenaceae.* After
Lord Robert Petre. Tender, evergreen
shrubs and climbers.
   volubilis vol-*ew*-bi-lis. Twining.
Purple Wreath, Sand Paper.

Petunia, pe-*tewn*-ee-a. *Solanaceae.*
From Brazilian *petun* (tobacco).
Annual, biennial and perennial herbs.
   axillaris, ax-il-*la*-ris. In the leaf axil.
Large White Petunia.
   x *hybrida,* hib-ri-da. Hybrid.
Petunia.
   integrifolia, in-teg-ri-*fol*-ee-a. Violet-
flowered Petunia.

Phacelia, fa-*sel*-ee-a.
Hydrophyllaceae. From Gk. *phakelos*
(bundle). Annual herbs.
   campanularia, kam-pan-ew-*la*-ree-a.
Campanula-like. California
Bluebell.
   tanacetifolia, tan-a-set-i-*fo*-lee-a.
Tanacetum-leaved. Fiddleneck.

Phalaris, fa-*la*-ris. *Gramineae.* From
Gk. for another grass. Annual and
perennial grasses.
   arundinacea, a-run-di-*na*-see-a.
Reed-like. Reed Canary Grass,

Gardener's Garters.
canariensis, ka-na-ree-*en*-sis. Of the
Canary Islands. Canary Grass.

Phalaris canariensis

Phaseolus, fa-*see*-o-lus. *Leguminosae.*
From Gk. *phaselos.* Annual herbs.
Tropical Bean.
   coccineus, kok-*kin*-ee-us. Scarlet.
   Scarlet Runner Bean.
   vulgaris, vul-*ga*-ris. Common.
   Kidney Bean, French Bean, Runner
   Bean.

Phellodendron, fe-lo-*den*-dron.
Rutaceae. From Gk. *phellos* (cork) and
dendron (tree), after the corky bark.
Deciduous trees.
   amurense, am-ew-*ren*-see. Of the
   Amur River region.

Philadelphus, fil-a-*del*-fus.
Hydrangaceae. Gk. name. Deciduous
shrubs. Mock Orange.
   coronarius, ko-ro-*na*-ree-us. Used in
   garlands.
   delavayi, de-la-*vay*-ee. After Delavay.

Philesia, fi-*leez*-ee-a. *Liliaceae.* From

Gk. *phileo* (love). Evergreen shrub.
   magellanica, ma-ge-*lan*-i-ka. Of the
   region of the Magellan Straits.

Phillyrea, fi-*li*-ree-a. *Oleaceae.* Gk.
name. Evergreen trees and shrubs.
   angustifolia, ang-gus-ti-*fo*-lee-a.
   Narrow-leaved. Mock Privet.
   decora, de-*ko*-ra. Beautiful.
   latifolia, la-ti-*fo*-lee-a. Broad-leaved.

Philodendron, fi-lo-*den*-dron.
Araceae. From Gk. *phileo* (love) and
dendron (tree). Tender, evergreen
shrubs and climbers.
   angustisectum, ang-gus-ti-*sek*-tum.
   With narrow divisions (leaves).
   bipennifolium, bi-pen-i-*fo*-lee-um.
   With bipinnate leaves. Horsehead
   Philodendron.
   domesticum, do-*mes*-ti-kum.
   Cultivated. Spade-leaf
   Philodendron.
   erubescens, e-roo-*bes*-enz.
   Blushing. Blushing Philodendron.
   hastatum, has-*ta*-tum. Spear-shaped.
   melanochrysum, me-la-*no*-kris-um.
   Black-gold..
   sagittifolium, sa-ji-ti-*fo*-lee-um.
   With arrow-shaped leaves.
   scandens, skan-denz. Climbing.
   Heart-leaf Philodendron.
   selloum, se-*lo*-um. After Sello.

Phlomis, *flo*-mis. *Labiatae.* Gk. for
another plant. Evergreen shrubs and
perennial herbs.
   cashmeriana, kash-me-ree-*a*-na. Of
   Kashmir.
   chrysophylla, kris-o-*fil*-la. Golden-
   leaved (when dried).
   fruticosa, froo-ti-*ko*-sa. Shrubby.
   Jerusalem Sage.
   italica, ee-*tal*-i-ka. Italian.
   russelliana, ru-sel-ee-*a*-na. After
   Russell.

Phlox, flox. *Polemoniaceae*. From Gk. *phlox* (flame). Annual and perennial herbs.

   adsurgens, ad-*sur*-genz. Erect.

   amoena, a-*mee*-na. Pleasant.

   bifida, bi-fi-da. Divided in two (petals).

   divaricata, di-va-ri-*ka*-ta. Spreading.

   douglasii, dug-*las*-ee-ee. After Douglas.

   drummondii, dru-*mond*-ee-ee. After Thomas Drummond. Annual Phlox.

   maculata, mak-ew-*la*-ta. Spotted (stems). Wild Sweet William.

   nana, na-na. Dwarf. Santa Fe Phlox.

   paniculata, pa-nik-ew-*la*-ta. With flowers in panicles.

   x *procumbens,* pro-*kum*-benz. Prostrate.

   stolonifera, sto-lo-*ni*-fe-ra. Bearing stolons. Creeping Phlox.

   subulata, sub-ew-*la*-ta. Awl-shaped (leaves). Moss Phlox.

Phoenix, *fee*-nix. *Palmae*. Gk. name. Tender, evergreen palms.

   canariensis, ka-na-ree-*en*-sis. Of the Canary Islands. Canary Island Date Palm.

   dactylifera, dak-ti-*li*-fe-ra. Finger-bearing. Date Palm.

   roebelinii, ro-be-*lin*-ee-ee. After M. Robelin. Miniature Date Palm. Pygmy Date Palm.

Phormium, *for*-mee-um. *Agavaceae*. From Gk. *phormion* (mat). Evergreen perennial herbs. Flax Lily.

   cookianum, kuk-ee-*a*-num. After Captain Cook. Mountain Flax.

   tenax, ten-ax. Tough. New Zealand Flax, New Zealand Hemp.

Photinia, fo-*tin*-ee-a. *Rosaceae*. From Gk. *photos* (light). Deciduous and evergreen shrubs and trees.

   arbutifolia, ar-bew-ti-*fo*-lee-a. Arbutus-leaved.

   beauverdiana, bo-ver-dee-*a*-na. After Gustave Beauverd.

   davidiana, da-vid-ee-*a*-na. After David.

   x *fraseri, fray*-za-ree. After the Fraser nurseries.

   glabra, glab-ra. Smooth (leaves).

   serrulata, se-ru-*la*-ta. With small teeth (leaves).

   villosa, vi-*lo*-sa. Softly hairy.

Phygelius, fi-*je*-lee-us. *Scrophulariaceae*. From Gk. *phyge* (flight) and *helios* (sun). Semi-hardy, semi-evergreen sub-shrubs.

   aequalis, ee-*kwa*-lis. Equal.

   capensis, ka-*pen*-sis. Of the Cape of Good Hope.

   x *rectus, rek*-tus. Upright.

Phyla, *fi*-la. *Verbenaceae*. From Gk. *phyla* (tribe). Perennial, creeping herb.

   canescens, ka-*nes*-enz. Greyish-white hairs. Carpet Grass.

   nodiflora, no-di-*flo*-ra. With flowers borne from the nodes. Frog Fruit.

Phyllocladus, fi-*lo*-kla-dus. *Podocarpaceae*. From Gk. *phyllon* (leaf) and *klados* (branch). Semi-hardy conifer.

   alpinus, al-*pie*-nus. Alpine. Alpine Celery Pine.

Phyllodoce, fi-*lo*-do-see. *Ericaceae*. After Phyllodoce, a sea nymph. Evergreen shrubs.

   breweri, broo-a-ree. After Brewer. Purple Heather.

   caerulea, see-*ru*-lee-a. Dark blue (corolla).

   empetriformis, em-pet-ri-*form*-is. *Empetrum*-like. Pink Mountain Heather.

nipponica, ni-*pon*-i-ka. Of Japan.

Phyllostachys, fi-*lo*-sta-kis.
Gramineae. From Gk. *phyllon* (leaf)
and *stachys* (spike), after the leafy
inflorescence. Bamboos. China.
  aurea, aw-ree-a. Golden (canes).
  Fishpole Bamboo.
  bambusoides, bam-bew-*soi*-deez.
  Bambusa-like. Giant Timber
  Bamboo.
  flexuosa, flex-ew-*o*-sa. Zig-zag
  (stems).
  nigra, nig-ra. Black. Black Bamboo.

Physalis, *fi*-sa-lis. *Solanaceae*. From
Gk. *physa* (bladder), after the bladder-
like fruits. Annual and perennial herbs.
Ground Cherry.
  alkekengi, al-ke-*ken*-jee. From
  Arabic *al kakendi.* Chinese Lantern.
  peruviana, pe-roo-vee-*a*-na. Of Peru.
  Cape Gooseberry.

Physocarpus, fi-so-*kar*-pus. *Rosaceae.*
From Gk. *physa* (bladder) and *karpon*
(fruit), after the inflated fruits.
Deciduous shrubs. Ninebark.
  amurensis, am-oor-*en*-sis. Of the
  Amur River region.
  opulifolius, op-ew-li-*fo*-lee-us. With
  leaves like *Viburnum opulus.*

Physoplexis, fi-so-*plex*-is.
Campanulaceae. From Gk. *physa*
(bladder) and *plexis* (plaiting).
Perennial herb.
  comosa, ko-*mo*-sa. Tufted. Devil's
  Claw.

Physostegia, fi-so-*stee*-gee-a.
Labiatae. From Gk. *physa* (bladder)
and *stege* (covering). Perennial herbs.
Obedient Plant.
  virginiana, vir-jin-ee-*a*-na. Of
  Virginia.

Phyteuma, fi-*tew*-ma.
Campanulaceae. Gk. name. Perennial
herbs. Horned Rampion.
  comosum, ko-*mo*-sum. Tufted.
  hemisphaericum, he-mis-*fer*-i-kum.
  Hemispherical (flower heads).
  orbiculare, or-bik-ew-*la*-ree.
  Orbicular. Roundheaded Rampion.
  spicatum, spee-*ka*-tum. With flowers
  in spikes. Spiked Rampion.

Phyteuma orbiculare

Phytolacca, fi-to-*la*-ka.
Phytolaccaceae. From Gk. *phyton*
(plant) and L. *lacca* (lac insect
Laccifer lacca from which dye is
obtained). Perennial herbs and evergreen shrubs.
  americana, a-me-ri-*ka*-na. Of
  America. Pokeweed.

Picea, *pi*-see-a. *Pinaceae.* From Gk.
pix (pitch). Pitch-producing pine.
Evergreen conifers. Spruce.
  abies, a-bee-ez. L. for fir (*Abies*).
  Norway Spruce.
  asperata, a-spe-*ra*-ta. Rough
  (foliage). Dragon Spruce.
  brachytyla, bra-kee-*ti*-la. With short

swellings. Sargent Spruce.

breweriana, broo-a-ree-*a*-na. After William Henry Brewer. Brewer's Spruce.

glauca, glow-ka. Glaucous (leaves). White Spruce.

likiangensis, li-kee-ang-*jen*-sis. Of Lakiang, Yunnan.

mariana, ma-ree-*a*-na. Of Maryland. Black Spruce.

omorika, o-*mo*-ri-ka. The native name. Serbian Spruce.

orientalis, o-ree-en-*ta*-lis. Eastern. Turkey, Caucasian Spruce.

pungens, pun-jenz. Sharp-pointed (leaves). Blue Spruce.

purpurea, pur-*pur*-ree-a. Purple (cones).

smithiana, smith-ee-*a*-na. After Sir James Edward Smith.

Picrasma, pik-*ras*-ma. *Simaroubaceae.* From Gk. *pikra* (bitter taste), bitter leaves and wood. Deciduous tree.

quassioides, kwa-see-*oi*-deez. Like *Quassia amara.*

Pieris, *pi*-e-ris. *Ericaceae.* From Gk. *Pierides* (Muses). Evergreen shrubs.

floribunda, flo-ri-*bun*-da. Profusely flowering. Fetter Bush.

formosa, for-*mo*-sa. Beautiful.

forrestii, fo-*rest*-ee-ee. After Forrest.

japonica, ja-*pon*-i-ka. Lily of the Valley Bush.

taiwanensis, tie-wan-*en*-sis. Of Taiwan.

Pilea, *pi*-lee-a. *Urticaceae.* From L. *pileus* (cap). Tender, trailing annuals and evergreen herbs.

cadierei, ka-dee-*e*-ree-ee. After R. P. Cadiere. Aluminium Plant.

involucrata, in-vo-loo-*kra*-ta. With an involucre. Friendship Plant.

microphylla, mik-ro-*fil*-la. Small-leaved. Artillery Plant, Pistol Plant.

nummulariifolia, num-ew-la-ree-i-*fo*-lee-a. With coin-shaped leaves.

repens, ree-penz. Creeping. Black-leaf Panamica.

spruceana, sproo-see-*a*-na. After Richard Spruce.

Pileostegia, pil-ee-o-*stee*-gee-a. *Hydrangeaceae.* From Gk. *pilos* (cap) and *stege* (covering). Evergreen climber.

viburnoides, vi-burn-*oi*-deez. *Viburnum*-like (flower heads).

Pimelea, pi-*me*-lee-a. *Thymelaeaceae.* From Gk. *pimele* (fat), after the oily seeds. Tender, evergreen shrubs.

ferruginea, fe-roo-*jin*-ee-a. Rusty.

prostrata, pros-*tra*-ta. Prostrate.

Pinguicula, pin-*gwi*-kew-la. *Lentibulariaceae.* From L. *pinguis* (fat), after the greasy appearance of the leaves. Carnivorous, perennial herbs. Butterwort.

caudata, kaw-*da*-ta. With a tail (spur).

grandiflora, grand-i-*flo*-ra. Large-flowered.

Pinguicula vulgaris

gypsicola, jip-*si*-ko-la. Lime-loving.
vulgaris, vul-*ga*-ris. Common.

Pinus sylvestris

Pinus, *pi*-nus. *Pinaceae.* The L. name.
Evergreen conifers. Pine.
    aristata, a-ris-*ta*-ta. Awned (the slen-
der cone bristles). Rocky Mountains
Bristlecone Pine.
    armandii, ar-*mond*-ee-ee. After
Armand David. Chinese White Pine.
    bungeana, bung-jee-*a*-na. After
Alexander von Bunge.
    cembra, sem-bra. Italian name.
Swiss Pine, Arolla Pine.
    contorta, kon-*tor*-ta. Twisted (young
shoots). Shore Pine.
    coulteri, kool-ta-ree. After Thomas
Coulter. Big-cone Pine.
    densiflora, den-si-*flo*-ra. Densely
flowered. Japanese Red Pine.
    halepensis, ha-le-*pen*-sis. Of Aleppo.
Aleppo Pine.
    heldreichii, hel-*driek*-ee-ee. After
Theodor von Heldreich. Bosnian
Pine.
    jeffreyi, jef-ree-ee. After John
Jeffrey.
    montezumae, mon-tee-*zoo*-mee.

After Montezuma. Montezuma Pine.
    muricata, mew-ri-*ka*-ta. Rough with
spines (cone). Bishop Pine.
California.
    nigra, nig-ra. Black (bark). Black
Pine.
    parviflora, par-vi-*flo*-ra. Small-flow-
ered. Japanese White Pine.
    pinea, pi-nee-a. L. for pine-nuts.
Stone Pine.
    ponderosa, pon-de-*ro*-sa. Heavy
(wood). Western Yellow Pine.
    rigida, ri-ji-da. Rigid (leaves).
Northern Pitch Pine.
    sylvestris, sil-*ves*-tris. Of woods.
Scots Pine.
    thunbergii, thun-*berg*-ee-ee. After
Thunberg. Japanese Black Pine.
    wallichiana, wo-lik-ee-*a*-na. After
Nathaniel Wallich. Himalayan Pine.

Piptanthus, pip-*tanth*-us,
Leguminosae. From Gk. *pipto* (fall)
and *anthos* (flower). Deciduous or
evergreen shrubs.
    nepalensis, ne-pa-*len*-sis. Of Nepal.
    tomentosus, to-men-*to*-sus. Hairy.

Pistacia, pis-*ta*-she-a. *Anacardiaceae.*
From Gk. *pistake* (pistachio nut).
Evergreen or deciduous trees or
shrubs.
    chinensis, chi-*nen*-sis. Of China.

Pistia, *pis*-tee-a. *Araceae.* From Gk.
pistos (water). Floating, tender perennial herb.
    stratiotes, stra-tee-*o*-teez. Gk. Water
Lettuce, Shell Flower.

Pittosporum, pi-*tos*-po-rum.
Pittosporaceae. From Gk. *pitta* (pitch)
and *sporum* (seed), the sticky seeds.
Tender and semi-hardy, evergreen trees
and shrubs.
    crassifolium, kras-i-*fo*-lee-um.

Thick-leaved. Evergreen
Pittosporum.
dallii, dal-ee-ee. After J. Dall.
eugenioides, ew-jeen-ee-*oi*-deez.
Eugenia-like. Lemonwood.
ralphii, ralf-ee-ee. After Dr Ralph.
tenuifolium, ten-ew-i-*fo*-lee-um.
With thin leaves.
tobira, to-*bi*-ra. The native name.
undulatum, un-dew-*la*-tum. Wavy-
edged (leaves). Orange Berry
Pittosporum.

Plagianthus, pla-jee-*anth*-us.
Malvaceae. From Gk. *plagios*
(oblique) and *anthos* (flower), after the
asymmetrical flowers. Evergreen or
deciduous shrubs and trees. Ribbon
Wood.
betulinus, bet-ew-*leen*-us. *Betula*-
like.
divaricatus, di-va-ri-*ka*-tus.
Spreading.
regius, ree-jee-us. Royal. Ribbon
Wood.

Platanus, *pla*-ta-nus. *Platanaceae.*
From Gk. *platanos.* Deciduous trees.
Plane.
x *acerifolia,* ay-se-ri-*fo*-lee-a. With
Acer-like leaves. London Plane.
orientalis, o-ree-en-*ta*-lis. Eastern.
Oriental Plane.
racemosa, ra-see-*mo*-sa. With flowers in racemes. California Sycamore.

Platycarya, pla-ti-*ka*-ree-a.
Juglandaceae. From Gk. *platys*
(broad) and *karyon* (nut). Deciduous
tree.
strobilacea, stro-bi-*la*-see-a. Cone-
like (fruit).

Platycerium, pla-ti-*se*-ree-um.
Polypodiaceae. From Gk. *platys*
(broad) and *keras* (horn) referring to

the flat, horn-like fronds. Tender
Ferns.
bifurcatum, bi-fur-*ka*-tum. Forked
into two. Elkshorn Fern.
grande, grand-ee. Large. Staghorn
Fern.

Platycodon, pla-ti-*ko*-don.
Campanulaceae. From Gk. *platys*
(broad) and *kodon* (bell), after the
corolla shape. Perennial herbs.
grandiflorus, grand-i-*flo*-rus. Large-
flowered. Balloon Flower.

Platystemon, pla-ti-*stee*-mon.
Papaveraceae. From Gk. *platys*
(broad) and *stemon* (stamen), after the
broad stamens. Annual herb.
californicus, kal-i-*forn*-i-kus. Of
California. Cream Cups, Californian
Poppy.

Plectranthus, plek-*tranth*-us.
Labiatae. From Gk. *plectron* (spur)
and *anthos* (flower). Tender, evergreen, trailing or bushy perennial
herbs.
australis, aw-*stra*-lis. Southern.
Swedish Ivy.
coleoides, ko-lee-*oi*-deez. *Coleus*-
like.
oertendahlii, ur-tan-*dal*-ee-ee. After
Oertendahl. Candle Plant.

Pleione, *ple*-o-nee. *Orchidaceae.* After
Pleione, wife of Atlas. Deciduous, cool
greenhouse orchids. Indian Crocus.
hookeriana huk-a-ree-*a*-na. After Sir
Joseph Hooker.
humilis, hu-mi-lis. Low-growing.
praecox, pree-kox. Early (flowering).

Pleiospilos, plee-*os*-pi-los. *Aizoaceae.*
From Gk. *pleios* (many) and *spilos*
(spot), after the spotted leaves. Tender
perennial succulents.
bolusii, bo-*lus*-ee-ee. After Harry

Bolus. Living Rock Cactus.

Plumbago, plum-*ba*-go.
Plumbaginaceae. From L. *plumbum*
(lead). Tender shrubs, perennials and
climbers. Leadwort.
    auriculata, aw-rik-ew-*la*-ta. With
    auricles (leaves). Cape Leadwort.
    caerulea, see-*ru*-lee-a. Dark blue.
    capensis, ka-*pen*-sis. Of the Cape of
    Good Hope.
    indica, in-di-ka. Of India. Scarlet
    Leadwort.

Plumeria, ploo-*me*-ree-a.
Apocynaceae. After Charles Plumier.
Tender, deciduous tree.
    alba, al-ba. White. West Indian
    Jasmine.
    obtusa, ob-*tew*-sa. Blunt.
    rubra, rub-ra. Red. Frangipani.

Podocarpus, pod-o-*kar*-pus.
Podocarpaceae. From Gk. *podos*
(foot) and *karpos* (fruit), after the
fleshy stalk of the fruit. Evergreen
conifers.
    alpinus, al-*pie*-nus. Alpine.
    Tasmanian Podocarp.
    andinus, an-*deen*-us. Of the Andes.
    Yacca Podocarp.
    macrophyllus, mak-ro-*fil*-lus. Large-
    leaved. Bigleaf Podocarp.
    nivalis, ni-*va*-lis. Growing near
    snow.
    salignus, sa-*lig*-nus. Willow-like
    (leaves). Willow Podocarp.

Podophyllum, pod-o-*fil*-lum.
Berberidaceae. From Gk. *anas* (duck),
podos (foot) and *phyllon* (leaf).
Rhizatomatous perennial herbs.
    hexandrum, hex-*an*-drum. With six
    stamens.
    peltatum, pel-*ta*-tum. Shield-shaped
    (leaves).Wild Mandrake.

Polemonium, po-li-*mo*-nee-um,
Polemoniaceae. From Gk. name *pole-
monion*. Annual and perennial herbs.
caeruleum, see-*ru*-lee-um. Dark blue.
Jacob's Ladder.
    caeruleum, see-*ru*-lee-um. Dark blue.
    carneum, kar-nee-um. Flesh-
    coloured.
    foliosissimum, fo-lee-o-*sis*-i-mum.
    Very leafy. Leafy Jacob's Ladder.
    pauciflorum, paw-si-*flo*-rum. Few-
    flowered.
    pulcherrimum, pul-*ke*-ri-mum. Very
    pretty.
    reptans, rep-tanz. Creeping. Greek
    Valerian.

Polemonium caeruleum

Polianthes, po-li-*anth*-eez. *Agavaceae*.
From Gk. *polios* (white) and *anthos*
(flower). Tender perennial herb.
    geminiflora, jem-in-ee-*flo*-ra.
    Having several flowers.
    tuberosa, tew-be-*ro*-sa. Tuberous.
    Tuberose.

Polygala, po-*li*-ga-la. *Polygalaceae*.
From Gk. *polys* (much) and *gala*
(milk). Herbs and shrubs. Milkworts

calcarea, kal-*sa*-ree-a. Growing on chalk.

chamaebuxus, ka-mee-*bux*-us. Dwarf *Buxus.*

grandiflora, gran-di-*flo*-ra. With large flowers.

vayredae, vay-*re*-dee. After Vayreda.

vulgaris, vul-*ga*-ris. Common. Milkwort.

Polygonatum multiflorum

Polygonatum, po-li-go-*na*-tum. *Liliaceae.* From Gk. *polys* (many) and *gony* (knee), after the jointed rhizomes. Perennial herbs. Solomon's Seal.

commutatum, kom-ew-*ta*-tum. Changeable.

hirtum, *hir*-tum. Hairy.

hookeri, *hu*-ka-ree. After Sir Joseph Hooker.

multiflorum, mul-ti-*flo*-rum. Many-flowered.

odoratum, o-do-*ra*-tum. Scented.

verticillatum, ver-ti-ki-*la*-tum. Whorled (leaves).

Polygonum, po-*li*-go-num. *Polygonaceae.* From Gk. *polys* (many) and *gony* (knee), after the jointed stems. Annual and perennial herbs and climbers.

affine, a-*fe*-nee. Related to.

amphibium, am-*fi*-bee-um. Growing in water or on land.

amplexicaule, am-plex-i-*kaw*-lee. With stem-clasping leaves.

aubertii, o-*bair*-tee-ee. After Aubert. Russian Vine.

baldschuanicum, bald-shoo-*an*-i-kum. Of Balzhuan. Mile-a-Minute Vine.

campanulatum, kam-pan-ew-*la*-tum. Bell-shaped (corolla). Lesser Knotweed.

capitatum, ka-pi-*ta*-tum. In a dense head (flowers).

multiflorum, mul-ti-*flo*-rum. Many-flowered.

sphaerostachyum, sfee-ro-*stak*-ee-um. With spherical flower heads.

tenuicaule, ten-ew-i-*kaw*-lee. Slender-stemmed.

Polygonum amphibium

Polypodium, po-li-*pod*-ee-um. *Polypodiaceae.* From Gk. *polys* (many) and *podos* (foot), after the branched rhizomes. Deciduous and evergreen ferns.

aureum, aw-ree-um. Golden (sori).
virgianum, ver-jin-ee-*a*-num. White.
American Wall Fern.
vulgare, vul-*ga*-ree. Common.
Common Polypody, Adders Fern.

Polyscias, po-*lis*-see-as. *Araliaceae.*
From Gk. *polys* (many) and *skias*
(umbel), after the abundant foliage.
Tender, evergreen trees and shrubs.
balfouriana, bal-for-ree-*a*-na. After
Sir Isaac Balfour.
filicifolia, fil-is-i-*fol*-ee-a. With fern-
like leaves. Angelica.
guilfoylei, gill-*foy*-lee-ee. After W.
R. Guilfoyle. Wild Coffee.

Polystichum, po-*li*-sti-kum.
Dryopteridaceae. From Gk. *polys*
(many) and *stichos* (row), after the
rows of sori. Evergreen and deciduous
ferns.
acrostichoides, a-kro-sti-*koi*-deez.
Acrostichum-like. Christmas Fern.
aculeatum, a-kew-lee-*a*-tum. Prickly.
munitum, mew-*ni*-tum. Armed
(teeth).
setiferum, se-*ti*-fe-rum. Bristly. Soft
Shield Fern.
tsu-simense, tsoo-see-*men*-see. Of
Tsu-shima, Japan.

Poncirus, pon-*si*-rus. *Rutaceae.* From
French *poncire* (citron). Deciduous
shrub.
trifoliata, tri-fo-lee-*a*-ta. With three-
leaves. Japanese Bitter Orange.

Pontederia, pon-te-*de*-ree-a.
Pontederiaceae. After Guilo
Pontedera. Deciduous, aquatic perennial herb.
cordata, kor-*da*-ta. Heart-shaped
(leaves). Pickerel Weed.

Populus alba

Populus, *po*-pu-lus. *Salicaceae.* L.
name. Deciduous trees. Poplar,
Cottonwood.
alba, al-ba. White (under the
leaves). White Poplar, Abele.
balsamifera, bal-sa-*mi*-fe-ra.
Balsam-bearing. Balsam Poplar.
x *berolinensis* be-ro-leen-*en*-sis. Of
Berlin.
canadensis, kan-a-*den*-sis. Of
Canada. Canadian Poplar.
candicans, kan-di-kanz. White
(under the leaves). Balm of Gilead
canescens, ka-*nes*-enz. Greyish-
whitehairs (leaves). Grey Poplar.
lasiocarpa, la-see-o-*kar*-pa. With
woolly fruits. Chinese Necklace
Poplar.
nigra, nig-ra. Black (bark). Black
Poplar.
tremula, trem-ew-la. Trembling
(leaves). Aspen, Quaking Aspen.
trichocarpa, tri-ko-*kar*-pa. With
hairy fruit. Black Cottonwood.

Portulaca, por-tew-*la*-sa.
Portulacaceae. L. name. Succulent

annuals and perennials.

grandiflora, grand-i-*flo*-ra. Large-flowered. Sun Plant, Rose Moss.

oleracea, o-le-*ra*-see-a. Vegetable-like. Purslane.

Potamogeton, po-ta-mo-*ge*-ton. *Potamogetonaceae.* From Gk. *potamos* (river) and *geiton* (neighbour). Deciduous, perennial, aquatic herbs.

lucens, lew-senz. Bright. Shining Pondweed.

crispus, kris-pus. Wavy (leaves). Curled Pondweed.

pectinatus, pek-ti-*na*-tus. Comb-like (leaves). Fennel Pondweed.

perfoliatus, per-fo-lee-*a*-tus. With the leaf surrounding the stem. Perfoliate Pondweed.

Potentilla, po-ten-*til*-la. *Rosaceae.* From L. *potens* (powerful), after its healing properties. Perennial herbs and deciduous shrubs.

alba, al-ba. White.

arbuscula, ar-*bus*-kew-la. Like a small tree.

Potentilla fruiticosa

argyrophylla, ar-ji-ro-*fil*-la. With silvery leaves

atrosanguinea, at-ro-sang-*gwin*-ee-a. Deep red.

aurea, aw-ree-a. Golden.

calabra, ka-*lab*-ra. Of Calabria.

crantzii, krantz-ee-ee. After H. J. N. von Crantz.

eriocarpa, e-ri-o-*kar*-pa. With woolly fruits.

fragiformis, fra-ji-*form*-is. Strawberry-like.

fruticosa, froo-ti-*ko*-sa. Shrubby. Shrubby Cinquefoil.

nepalensis, ne-pa-*len*-sis. Of Nepal.

nitida, ni-ti-da. Shining.

recta, rek-ta. Erect.

Pratia, *pra*-tee-a. *Campanulaceae.* After Ch. L. Prat-Bernon. Evergreen, perennial herb.

angulata, ang-gew-*la*-ta. Angled.

pedunculata, ped-unk-ew-*la*-ta. With a flower stalk.

Primula, *prim*-ew-la. *Primulaceae.* From L. *primus* (first), after its early flowering. Annual, biennial and perennial herbs.

acaulis, a-*kaw*-lis. Stemless.

allionii, a-lee-*o*-nee-ee. After Carlo Allioni.

alpicola, al-*pi*-ko-la. Growing on mountains.

amoena, a-*mee*-na. Pleasant.

aurantiaca, aw-ran-tee-*a*-ka. Orange (flowers).

auricula, aw-*rik*-ew-la. Ear-like.

bulleyana, bu-lee-*a*-na. After A. K. Bulley.

burmanica, bur-*man*-i-ka. Of Burma.

capitata, ka-pi-*ta*-ta. In a dense head (flowers).

chionantha, kee-on-*anth*-a. With snow-white flowers.

clarkei, klark-ee-ee. After C. B. Clarke.

cockburniana, ko-burn-ee-*a*-na. After H. Cockburn.

cortusoides, kor-tew-*soi*-deez. *Cortusa*-like.

denticulata, den-tik-ew-*la*-ta. Slightly toothed. Drumstick Primula.

edgeworthii, ej-*werth*-ee-ee. After Edgeworth.

elatior, e-*la*-tee-or. Taller. Oxlip.

florindae, flo-*rin*-dee. After Kingdon-Ward's wife Florinda.

frondosa, fron-*do*-sa. Leafy.

helodoxa, he-lo-*dox*-a. Glory of the marsh.

hirsuta, hir-*soo*-ta. Hairy.

japonica, ja-*pon*-i-ka. Of Japan.

malacoides, ma-la-*koi*-deez. Mallow-like. Fairy Primrose, Baby Primrose.

marginata, mar-ji-*na*-ta. Margined (leaves).

minima, mi-ni-ma. Smallest.

nutans, new-tanz. Nodding (flowers).

obconica, ob-*ko*-ni-ka. Like an inverted cone (calyx). German Primrose.

polyantha, po-lee-*anth*-a. Many-flowered. Polyanthus.

polyneura, po-lee-*new*-ra. With many veins.

prolifera, pro-*li*-fe-ra. Proliferous.

reptans, rep-tanz. Creeping.

pulverulenta, pul-ve-ru-*len*-ta. Mealy.

rosea, ro-see-a. Rose-coloured.

secundiflora, se-kun-di-*flo*-ra. With flowers on one side of the stalk.

sieboldii, see-*bold*-ee-ee. After Siebold.

sikkimensis, si-kim-*en*-sis. Of Sikkim.

sinensis, si-*nen*-sis. Of China.

spectabilis, spek-*ta*-bi-lis. Spectacular.

veris, ve-ris. Of spring. Cowslip.

vialii, vee-*al*-ee-ee. After Père Vial.

vulgaris, vul-*ga*-ris. Common. Primrose.

Proboscidea, pro-bos-*si*-dee-a. P*edaliaceae.* From Gk. *proboskis* (elephant's trunk), after the long, curved fruit beak. Annual and perennial semi-hardy herbs.

fragrans, fra-granz. Fragrant.

Prostanthera, pros-tanth-*e*-ra. *Labiatae.* From Gk. *prosthema* (appendage) and *anthera* (anther). Semi-hardy, evergreen shrubs. Australian Mint Bush.

ovalifolia, o-va-li-*fo*-lee-a. With oval leaves.

rotundifolia, ro-tun-di-*fo*-lee-a. Round-leaved. Mint Bush.

Protea, *pro*-tee-a. *Proteaceae.* After Proteus, a Gk. sea god. Tender, evergreen shrubs and trees.

barbigera, bar-*bi*-je-ra. Bearded.

cynaroides, si-na-*roi*-deez. *Cynara*-like. King Protea.

eximea, ex-*i*-mee-a. Distinguished.

Primula vulgaris

grandiceps, grand-i-seps. Large-headed.

magnifica, mag-*ni*-fi-ka. Magnificent.

neriifolia, nee-ree-i-*fo*-lee-a. With *Nerium*-like leaves.

repens, ree-penz. Creeping.

Prunella, proo-*nel*-a. *Labiatae.* From German *Braune* (quinsy), after its alleged healing properties. Perennial herbs. Self-heal.

grandiflora, grand-i-*flo*-ra. Large-flowered.

x *webbiana,* web-ee-*a*-na. After Webb.

Prunus, *proo*-nus. *Rosaceae.* L. for plum tree. Deciduous or evergreen shrubs and trees. Plum, Cherry, Peach, Almond, Apricot.

americana, a-me-ri-*ka*-na. Of America. Wild Plum.

armeniaca, ar-men-ee-*a*-ka. Of Armenia. Apricot.

avium, a-vee-um. Of birds. Bird Cherry, Sweet Cherry.

x *blireana,* bli-ree-*a*-na. Of Bléré, France.

cerasifera, se-ra-*si*-fe-ra. Cherry-bearing. Cherry Plum, Myrobalan.

cerasus, se-ra-sus. L. name for cherry. Sour Cherry.

davidiana, da-vid-ee-*a*-na. After David. David's Peach.

domestica, do-*mes*-ti-ka. Cultivated. Common Plum.

dulcis, dul-sis. Sweet. Almond.

glandulosa, glan-dew-*lo*-sa. Glandular. Dwarf Flowering Almond.

incisa, in-*see*-sa. Deeply cut (leaves). Fuji Cherry.

laurocerasus, law-ro-*se*-ra-sus. Cherry Laurel, Laurel Cherry.

lusitanica, loo-si-*ta*-ni-ka. Of

Prunus cerasus

Portugal. Portuguese Laurel Cherry.

mume, mew-mee. From *ume* the Japanese name. Japanese Apricot.

persica, per-si-ka. Of Persia. Peach.

pumila, pew-mi-la. Dwarf. Sand Cherry.

serrula, se-ru-la. Saw-toothed (leaves). Birch-bark Tree.

serrulata, se-ru-*la*-ta. With small teeth (leaves). Oriental Cherry.

spinosa, spi-*no*-sa. Spiny. Sloe, Blackthorn.

subhirtella, sub-hir-*tel*-la. Somewhat hairy. Winter Flowering Cherry.

tenella, te-*nel*-la. Dainty. Dwarf Russian Almond.

texana, tex-*a*-na. Of Texas. Peach Bush.

triloba, tri-*lo*-ba. Three-lobed (leaves). Flowering Almond.

Pseuderanthemum, soo-de-*ranth*-e-mum. *Acanthaceae.* From Gk. *pseudo* (false) and *Eranthemum.* Tender evergreen perennials and shrubs.

alatum, a-*la*-tum. Winged. Chocolate Plant.

atropurpureum, at-ro-pur-*pur*-ree-um. Deep purple (leaves).

reticulatum, re-tik-ew-*la*-tum. Net-veined (leaves).

Pseudopanax, soo-do-*pan*-ax. *Araliaceae.* From Gk. *pseudo* (false) and *Panax.* Semi-hardy, evergreen trees or shrubs.
  arboreus, ar-*bo*-ree-us. Tree-like.
  crassifolius, kras-i-*fo*-lee-us. Thick-leaved.
  davidii, da-*vid*-ee-ee. After David.
  ferox, fe-rox. Spiny (leaves). Toothed Lancewood.
  laetus, *lee*-tus. Bright.

Pseudotsuga, soo-do-soo-ga. *Pinaceae.* From Gk. *pseudo* (false) and *Tsuga.* Evergreen conifer.
  menziesii, men-*zeez*-ee-ee. After Menzies. Green Douglas Fir.

Pseudowintera, soo-do-win-*te*-ra. *Winteraceae.* From Gk. *pseudo* (false) and *Wintera.* Evergreen trees and shrubs.
  axillaris, ax-il-*la*-ris. In the leaf axil.
  colorata, ko-lo-*ra*-ta. Coloured (leaves).

Psylliostachys, si-lee-*o*-sta-kis. *Plumbaginaceae.* From Gk. *psyllion* (plantain) and *stachys* (spike). Annuals, perennials and sub-shrubs.
  spicata, spi-*ka*-ta. With flowers in spikes.
  suworowii, soo-vo-*rov*-ee-ee. After Ivan Petrowitch Suworow.

Ptelea, *tel*-ee-a. *Rutaceae.* Gk. for elm. Deciduous trees and shrubs.
  trifoliata, tri-fo-lee-*a*-ta. With three leaves. Hop Tree, Water Ash.

Pteris, *te*-ris. *Pteridaceae.* From Gk. *pteris* (fern). Tender ferns.
  cretica, *kree*-ti-ka. Of Crete. Cretan Brake.

multifida, mul-ti-*fi*-da. Divided many times.
quadriaurita, kwod-ree-aw-*ree*-ta. Four-eared.
tremula, trem-ew-la. Trembling. Australian Bracken
tripartita, tri-*part*-ee-ta. In three parts. Giant Bracken.

Pterocarya, te-ro-*ka*-ree-a. *Juglandaceae.* From Gk. *pteron* (wing) and *karyon* (nut), after the winged fruit. Deciduous trees. Wing-nut.
  fraxinifolia, frax-i-ni-*fo*-lee-a. *Fraxinus*-leaved. Caucasian Wing-nut.
  x *rehderiana,* ree-da-ree-*a*-na. After Rehder.

Pterocephalus, te-ro-*sef*-a-lus. *Dipsacaceae.* From Gk. *pteron* (wing) and *kephale* (head), after the 'feather' covered fruiting head. Annual and Perennial herbs and sub-shrubs.
  perennis, pe-*ren*-is. Perennial.

Pulmonaria, pul-mon-*a*-ree-a. *Boraginaceae.* From L. *pulmo* (lung), the leaves were used to treat diseases of the lungs. Herbaceous perennials. Lungwort.
  affinis, a-*fee*-nis. Related to.
  angustifolia, ang-gus-ti-*fo*-lee-a. Narrow-leaved.
  longifolia, long-i-*fo*-lee-a. Long-leaved.
  officinalis, o-fi-si-*na*-lis. Sold in shops. Jerusalem Sage.
  saccharata, sa-ka-*ra*-ta. Sugar-coated (leaves).

Pulsatilla, pul-sa-*til*-la. *Ranunculaceae.* From L. *pulso* (strike). Perennial herbs. Pasque Flower.
  alba, al-ba. White.
  alpina, al-*pie*-na. Alpine. Alpine Pasque Flower.

halleri, hal-a-ree. After Albrecht von
Haller.
patens, pa-tenz. Spreading. Eastern
Pasque Flower.
vernalis, ver-*na*-lis. Of spring.
vulgaris, vul-*ga*-ris. Common
Pasque Flower.

Punica, *pew*-ni-ka. *Punicaceae.* L.
name. Semi-hardy evergreen trees and
shrubs.
 granatum, gra-*na*-tum. Many-seed-
 ed. Pomegranate.

Puschkinia, push-*kin*-ee-a. *Liliaceae.*
After Count Mussin-Puschkin. Dwarf,
bulbous perennials.
 scilloides, sil-*loi*-deez. *Scilla*-like.

Pyracantha, pi-ra-*kanth*-a. *Rosaceae.*
From Gk. *pyr* (fire) and *akantha*
(thorn), after the spiny shoots and red
berries. Evergreen shrubs. Firethorn.
 angustifolia, ang-gus-ti-*fo*-lee-a.
 Narrow-leaved.
 atalantioides, a-ta-lan-tee-*oi*-deez.
 Atalantia-like. (Rutaceae).
 coccinea, kok-*kin*-ee-a. Scarlet
 (fruits). Pyracanth, Firethorn.
 crenulata, kren-ew-*la*-ta. Scalloped.
 Nepalese White Thorn.
 rogersiana, ro-jerz-ee-*a*-na. After G.
 L. Coltman-Rogers

Pyrola, *pi*-ro-la. *Pyrolaceae.*
Diminutive of *Pyrus.* Evergreen,
perennial herbs. Wintergreen, Shinleaf.

Pyrola media

 asarifolia, a-sa-ri-*fo*-lee-a. *Asarum*-
 leaved.
 chlorantha, klo-*ran*-tha. With green
 flowers. Green-flowered Wintergreen.
 elliptica, e-*lip*-ti-ka. Elliptic (leaves).
 media, me-dee-a. Intermediate.
 picta, pik-ta. Painted. White-veined
 Wintergreen.
 rotundifolia ro-tun-di-*fo*-lee-a.
 Round-leaved.

Pyrus, *pi*-rus. *Rosaceae.* L. name.
Deciduous trees. Pear.
 calleryana, ka-le-ree-*al*-na. After J.
 Callery. Callery Pear.
 communis, kom-*ew*-nis. Common.
 Common Pear.
 salicifolia, sa-li-si-*fo*-lee-a. *Salix*-
 leaved.

Q

Quercus, *kwer*-kus. *Fagaceae.* L.
name. Deciduous and evergreen trees.
Oak.

   acutissima, a-kew-*tis*-i-ma. Very
sharply pointed. Sawthorn Oak.
   agrifolia, ag-ri-*fo*-lee-a. With spiny
leaves. California Live Oak.
   alba, al-ba. White. White Oak.
   alnifolia, al-ni-*fo*-lee-a. *Alnus*-
leaved. Golden Oak.
   canariensis, ka-na-ree-*en*-sis. Of the
Canary Islands. Algerian Oak.
   castaneifolia, kas-tan-ee-i-*fo*-lee-a.
Castanea-leaved.
   cerris, se-ris. L. name. Turkey Oak.
   coccifera, kok-*kif*-e-ra. Berry-bearing. Grain Oak.
   coccinea, kok-*kin*-ee-a. Scarlet.
Scarlet Oak.
   dentata, den-*ta*-ta. Toothed.
Japanese Emperor Oak.
   ilex, ee-lex. L. name. Holm Oak,
Evergreen Oak.
   laurifolia, law-ri-*fo*-lee-a. Bay-
leaved. Laurel Oak.
   libani, li-ba-nee. Of Lebanon.
Lebanon Oak.
   macranthera, ma-*kranth*-e-ra. With
large anthers. Caucasian Oak.
   macrocarpa, mak-ro-*kar*-pa. Large-
fruited. Burr Oak.
   macrolepis, mak-ro-*lep*-is. With
large scales.
   marilandica, ma-ri-*land*-i-ka. Of
Maryland. Blackjack Oak.
   myrsinifolia, mur-si-ni-*fo*-lee-a.
Myrsine-leaved.

Quercus robur

   nigra, nig-ra. Black. Water Oak.
   palustris, pa-*lus*-tris. Of swamps.
Pin Oak.
   petraea, pe-*tree*-a. Of rocky places.
   phellos, fel-os. Willow-leaved.
Willow Oak.
   pontica, pon-ti-ka. Of the shore of
the Black Sea. Armenian Oak.
   pyrenaica, pi-ren-*ee*-i-ka. Of the
Pyrenees. Spanish Oak.
   robur, ro-bur. L. name. Common
Oak, English Oak.
   rubra, rub-ra. Red. Red Oak.
   suber, soo-ber. L. name. Cork Oak.
   x *turneri,* turn-a-ree. After Spencer
Turner.
   velutina, vel-ew-*teen*-a. Velvety.
Black Oak.
   wislizenii, wiz-li-*zen*-ee-ee. After A.
Wislizenius. Interior Live Oak.

R

Ramonda, ra-*mon*-da. *Gesneriaceae.*
After Louis Ramond. Evergreen,
perennial herbs.
> *myconi,* mi-*ko*-nee. After Franciso
> Mico.
> *nathaliae,* na-*ta*-lee-ee. After Queen
> Nathalia, wife of King Milan.
> *pyrenaica,* pi-ren-*ee*-i-ka. Of the
> Pyrenees.
> *serbica, ser*-bi-ka. Of Serbia.

Ranunculus, ra-*nun*-kew-lus.
Ranunculaceae. The L. name from
rana (frog). Annual, perennial and
aquatic herbs. Buttercup, Crowfoot.
> *aconitifolius,* a-kon-ee-ti-*fo*-lee-us.
> *Aconitum*-leaved. White Bachelors
> Buttons.
> *acris, a*-kris. Sharp-tasting. Meadow
> Buttercup.
> *alpestris,* al-*pes*-tris. Of the lower
> mountains.
> *amplexicaulis,* am-plex-i-*kaw*-lis.
> With leaves clasping the stem.
> *aquatilis,* a-*kwa*-ti-lis. Growing in
> water. Water Crowfoot.
> *asiaticus,* a-see-*a*-ti-kus. Asian.

Ranunculus acris

Persian Buttercup.
> *bullatus,* bu-*la*-tus. With puckered
> leaves.
> *calandrinioides,* ka-lan-dree-nee-*oi*-
> deez. *Calandrinia*-like .
> *ficaria,* fi-*ka*-ree-a. *Ficus*-like.
> Lesser Celandine.
> *glacialis,* gla-see-*a*-lis. Of icy
> regions..
> *gramineus,* gra-*min*-ee-us. Grass-like
> (leaves).
> *lingua, ling*-wa. Tongue-like
> (leaves). Greater Spearwort.
> *montana,* mon-*ta*-na. Of mountains.
> *pyrenaeus,* pi-ree-*nee*-us. Of the
> Pyrenees.
> *repens, ree*-penz. Creeping.
> Creeping Buttercup.
> *speciosus,* spes-ee-*o*-sus. Showy.

Raoulia, *rowl*-ee-a. *Compositae.* After
Edward Raoul. Evergreen perennial
herbs and sub-shrubs.
> *australis,* aw-*stra*-lis. Southern.
> *haastii, harst*-ee-ee. After Sir
> Johann von Haast.
> *hookeri,* huk-a-ree. After Sir Joseph
> Hooker.
> *tenuicaulis,* ten-ew-i-*kaw*-lis.
> Slender-stemmed.

Raphanus, *ra*-fa-nus. *Cruciferae.* L.
name. Biennial or perennial herbs.
> *caudatus,* kaw-*da*-tus. With a tail.
> *sativus,* sa-*teev*-us. Cultivated.
> Radish.

Rebutia, re-*bew*-shee-a. *Cactaceae.*
After P. Rebut. Cactus.
> *aureiflora,* aw-ree-i-*flo*-ra. Golden-
> flowered.

minuscula, mi-*nus*-kew-la. Rather
small. Red Crown Cactus.
pygmaea, pig-*mee*-a. Dwarf.
senilis, se-*nee*-lis. An old man.
spegazziniana, spe-ga-zeen-ee-*a*-na.
After Professor Carlos Spegazzini.
violaciflora, vie-o-la-si-*flo*-ra. With
violet flowers.

Rehderodendron, ree-da-ro-*den*-dron.
Styracaceae. After Alfred Rehder and
Gk. *dendron* (tree). Deciduous trees
and shrubs.
   macrocarpum, mak-ro-*kar*-pum.
   Large-fruited.

Rehmannia, ree-*man*-ee-a.
Gesneriaceae. After Joseph Rehmann.
Tender to semi-hardy perennial herbs.
   elata, e-*la*-ta. Tall.
   glutinosa, gloo-ti-*no*-sa. Sticky.

Reinwardtia, rien-*wardt*-ee-a.
Linaceae. After Caspar Reinwardt.
Tender, evergreen sub-shrubs.
   indica, in-di-ka. Of India.Yellow
   Flax.

Reseda, re-*se*-da. *Resedaceae.* From
L. *resedo* (heal), after its alleged healing properties. Annual and biennial
herbs.
   alba, al-ba. White. Wild Mignonette.
   odorata, o-do-*ra*-ta. Fragrant.
   Mignonette.

Rhamnus, *ram*-nus. *Rhamnaceae.* Gk.
name. Deciduous or evergreen trees
and shrubs.
   alaternus, a-la-*tern*-us. L. name.
   cathartica, ka-*thar*-ti-ka. Purging.
   Buckthorn.
   croceus, kro-see-us. Saffron yellow.
   Redberry.
   frangula, frang-gew-la. L. name.
   Alder Buckthorn.

Rhamnus frangula

Rhaphiolepis, raf-ee-o-*lep*-is.
Rosaceae. From Gk. *rhaphis* (needle)
and *lepis* (scale), after its needle-like
bracts. Semi-hardy, evergreen shrubs.
   x *delacourii,* de-la-*koor*-ee-ee. After
   M. Delacour.
   indica, in-di-ka. Of India. Indian
   Hawthorn.
   umbellata, um-bel-*a*-ta. Umbelled.

Rhapis, *ra*-pis. *Palmae.* From Gk.
rhapis (needle). Tender, evergreen fan
palms.
   excelsa, ex-*sel*-sa. Tall. Ground
   Rattan.
   humilis, hu-mi-lis. Low-growing.
   Slender Lady Palm.

Rheum, *ree*-um. *Polygonaceae.* From
Gk. *rheon* (rhubarb). Perennial herbs.
   alexandrae, a-lex-*an*-dree. After
   Queen Alexandra, wife of Edward VII.
   nobile, no-bi-lee. Notable.
   palmatum, pal-*ma*-tum. Hand-like.

Rhipsalidopsis, rip-sa-li-*dop*-sis,
Cactaceae. From *Rhipsalis* and Gk. -
opsis (resemblance).

gaertneri, gert-na-ree. After J.
Gartner.
rosea, ro-see-a. Rose-coloured
(flowers).

Rhipsalis, *rip*-sa-lis. *Cactaceae*. From
Gk. *rhips* (wicker-work), after the
intertwining shoots.
capilliformis, ka-pi-li-*form*-is.
Thread-like (stems)
cereuscula, see-ree-*us*-kew-la. Like
a small *Cereus*.
crispata, kris-*pa*-ta. Wavy-edged
(stems).
paradoxa, pa-ra-*dox*-a. Unusual.
Chain Cactus.
warmingiana, war-ming-gee-*a*-na.
After Professor Johannes Warming.

Rhodochiton, ro-*do*-ki-ton.
Scrophulariaceae. From Gk. *rhodo-*
(red) and *chiton* (cloak). Tender, evergreen climber.
volubile, vol-*ew*-bi-lee. Twining.
Purple Bell Vine.

Rhododendron, ro-do-*den*-dron.
Ericaceae. From Gk. *rhodo-* (red) and
dendron (tree). Evergreen or deciduous
trees and shrubs.
aberconwayi, a-ba-*kon*-way-ee. After
Lord Aberconway.
albrechtii, al-*brekt*-ee-ee. After Dr
Albrecht.
arborescens, ar-bo-*res*-enz.
Becoming tree-like.
arboreum, ar-*bor*-ee-um. Tree-like.
atlanticum, at-*lan*-ti-kum. Of the
Atlantic coast.
augustinii, aw-gus-*tin*-ee-ee. After
Augustine Henry.
auriculatum, aw-rik-ew-*la*-tum.
Auricled (leaves).
barbatum, bar-*ba*-tum. Bearded
(shoots of some forms).
bureavii, bew-*reev*-ee-ee. After

Edouard Bureau.
calendulaceum, ka-len-dew-*la*-see-
um. *Calendula*-like (flower colour).
callimorphum, kal-i-*mor*-fum.
Beautifully shaped.
calostrotum, kal-os-*tro*-tum. With a
beautiful covering.
caloxanthum, ka-lox-*anth*-um. A
beautiful yellow.
campylocarpum, kam-pi-lo-*kar*-pum.
With a curved fruit.
campylogynum, kam-pi-*lo*-ji-num.
With a curved ovary.
catawbiense, ka-taw-bee-*en*-see.
From near the Catawba River.
chasmanthum, kas-*manth*-um. With
gaping flowers.
ciliatum, si-lee-*a*-tum. Fringed with
hairs (leaves).
cinnabarinum, si-na-ba-*reen*-um.
Cinnabar red.
concinnum, kon-*sin*-um. Elegant.
crassum, kras-um. Thick (leaves).
decorum, de-*ko*-rum. Beautiful.
discolor, dis-ko-lor. Two-coloured
(leaves).
edgeworthii, ej-*werth*-ee-ee. After
Edgeworth.
falconeri, fol-*kon*-a-ree. After Hugh
Falconer.
fargesii, far-*jee*-zee-ee. After Farges.
fastigiatum, fa-stij-ee-*a*-tum. With
upright branches.
ferrugineum, fe-roo-*jin*-ee-um.
Rusty.
fictolacteum, fik-to-*lak*-tee-um.
False *R. lacteum*.
forrestii, fo-*rest*-ee-ee. After Forrest.
fortunei, for-*tewn*-ee-ee. After
Fortune.
fulvum, ful-vum. Tawny (lower leaf
surface).
glaucophyllum, glow-ko-*fil*-lum.
With glaucous leaves (undersides).
haematodes, hee-ma-*to*-deez. Blood-
red.

hanceanum, hans-ee-*a*-num. After Henry Fletcher Hance.

hippophaeoides, hi-po-fee-*oi*-deez. *Hippophae*-like.

hirsutum, hir-*soo*-tum. Hairy.

impeditum, im-pe-*dee*-tum. Tangled.

imperator, im-*pe*-ra-tor. Emperor.

indicum, in-di-kum. Of India.

insigne, in-*sig*-nee. Distinguished.

kaempferi, kempf-e-ree. After Engelbert Kaempfer.

kiusianum, kee-oo-see-*a*-num. Of Kyushu, Japan.

lepidostylum, le-pi-do-*sti*-lum. With a scaly style.

leucaspis, loo-*kas*-pis. A white shield.

lutescens, loo-*tes*-enz. Yellowish.

luteum, loo-tee-um. Yellow.

macabeanum, ma-ka-bee-*a*-num. After Mr McCabe.

maddenii, ma-*den*-ee-ee. After Major Madden.

moupinense, moo-pin-*en*-see. Of Moupin.

mucronulatum, mew-kron-ew-*la*-tum. With a short point (leaves).

neriiflorum, ne-ree-i-*flo*-rum. *Nerium*-flowered.

obtusum, ob-*tew*-sum. Blunt (leaves).

occidentale, ok-si-den-*ta*-lee. Western.

orbiculare, or-bik-ew-*la*-ree. Rounded (leaves).

oreotrephes, o-ree-*o*-tre-feez. Growing on mountains.

pemakoense, pe-ma-ko-*en*-see. Of Pemako, Tibet.

polycladum, po-lee-*klad*-um. With many branches.

ponticum, pon-ti-kum. Of Pontus (Turkey).

pruniflorum, proon-i-*flo*-rum. *Prunus*-flowered.

pseudochrysanthum, soo-do-kris-*anth*-um. False *R. chrysanthum.*

quinquefolium, kwin-kwee-*fo*-lee-um. With five leaves.

racemosum, ra-see-*mo*-sum. With flowers in racemes.

radicans, ra-di-kanz. With rooting stems.

repens, ree-pens. Creeping.

reticulatum, re-tik-ew-*la*-tum. Net-veined (lower leaf surface).

rex, rex. King.

rubiginosum, roo-bi-ji-*no*-sum. Rusty (lower leaf surface).

russatum, rus-*a*-tum. Russet (lower leaf surface).

sargentianum, sar-jent-ee-*a*-num. After Sargent.

Scintillans, sin-ti-lanz. Gleaming.

schlippenbachii, shlip-en-*bark*-ee-ee. After Baron Schlippenbach.

simsii, sim-zee-ee. After John Sims.

souliei, soo-lee-ee. After Jean Soulie.

sutchuenense, such-wen-*en*-see. Of Szechwan.

thomsonii, tom-*son*-ee-ee. After Thomson.

uniflorum, ew-ni-*flo*-rum. With one flower.

valentinianum, va-len-tin-ee-*a*-num. After Père S. P. Valentin.

vaseyi, vay-zee-ee. After George R. Vasey.

viscosum, vis-*ko*-sum. Sticky (flowers).

wardii, ward-ee-ee. After Kingdon Ward.

williamsianum, wil-yam-zee-*a*-num. After J. C. Williams.

xanthocodon, zanth-o-*ko*-don. A yellow bell.

yakushimanum, ya-koo-shee-*ma*-num. Of Yakushima.

yedoense, ye-do-*en*-see. Of Yedo (Tokyo).

yunnanense, yoo-nan-*en*-see. Of Yunnan.

Rhoicissus, ro-i-*sis*-us. *Vitaceae*. From
L. *rhoicus* (*Rhus*) and *Cissus*. Tender,
evergreen climbers.
  capensis, ka-*pen*-sis. Of the Cape of
  Good Hope.

Rhombophyllum, rom-bo-*fil*-lum.
Aizoaceae. From Gk. *rhombos* (rhom-
bus) and *phyllon* (leaf), after the leaf
shape. Tender perennial succulents.
  nelii, nel-ee-ee. After G. C. Nel.
  Elk's Horns.
  rhomboideum, rom-*boi*-dee-um.
  Diamond-shaped (leaves).

Rhus, *rus*. *Anacardiaceae*. L. name of
R. coriaria. Deciduous or evergreen
trees, shrubs and climbers. Sumac.
  aromatica, a-ro-*ma*-ti-ka. Fragrant.
  Lemon Sumac.
  copallina, ko-pal-*le*-na. Resinous.
  Dwarf Sumac.
  cotinoides, ko-*te*-noi-deez. Cotinus-
  like.
  cotinus, ko-*te*-nus. Old Gk. name.
  Smoke Tree.
  glabra, glab-ra. Smooth. Smooth
  Sumach.
  typhina, tie-*fee*-na. *Typha*-like.
  Staghorn Sumac.

Ribes, *rie*-beez. *Grossulariaceae*.
From Arabic *ribas* (acid-tasting), after
the fruit. Deciduous and evergreen
shrubs. Currant, Gooseberry.
  alpinum, al-*pie*-num. Alpine.
  Mountain Currant.
  aureum, aw-ree-um. Golden. Golden
  Currant.
  laurifolium, law-ri-*fo*-lee-um.
  Laurus-leaved.
  nigrum, nig-rum. Black (fruit).
  Blackcurrant.
  odoratum, o-do-*ra*-tum. Fragrant
  (flowers). Clove Currant.
  rubrum, rub-rum. Red. Redcurrant.

Ribes alpinum

  sanguineum, sang-*gwin*-ee-um.
  Blood-red. Winter Currant.
  speciosum, spes-ee-*o*-sum. Showy.
  Fuschia-flowered Gooseberry.
  uva-crispa, oo-va-*kris*-pa. Crisp
  grape. Gooseberry.

Richea, *reesh*-ee-a. *Epacridaceae*.
After M. Riche, Evergreen shrub.
  scoparia, sko-*pa*-ree-a. Broom-like.

Ricinus, *ri*-si-nus. *Euphorbiaceae*.
From L. *ricinus* (tick), after the tick-
like seeds. Semi-hardy evergreen
shrub.
  communis, kom-*ew*-nis. Common.
  Castor Oil Plant.

Robinia, ro-*bin*-ee-a. *Leguminosae*.
After Jean Robin. Deciduous trees and
shrubs.
  ambigua, am-*big*-ew-a. Doubtful.
  hispida, his-pi-da. Bristly (shoots).
  Rose Acacia, Moss Locust.
  kelseyi, kel-see-ee. After Mr Harlan
  P. Kelsey.
  pseudacacia, sood-a-*kay*-she-a.
  False *Acacia*. Black Locust, Yellow
  Locust.

Rochea, *rosh*-ee-a. *Crassulaceae.*
After Daniel de la Roche. Tender evergreen succulent shrubs.
  coccinea, kok-*kin*-ee-a. Scarlet.

Rodgersia, ro-*jerz*-ee-a.
Saxifragaceae. After Rear Admiral
John Rodgers. Perennial herbs.
  aesculifolia, ee-skew-li-*fo*-lee-a.
  With *Aesculus*-like leaves.
  pinnata, pi-*na*-ta. Pinnate.
  podophylla, po-do-*fil*-a. With stoutly
  stalked leaves.
  sambucifolia, sam-bew-ki-*fo*-lee-a.
  Sambucus-leaved.

Romneya, rom-*nee*-a. *Papaveraceae.*
After Dr Thomas Robinson.
Perennial herbs and deciduous
sub-shrubs. Californian Tree Poppy.
  coulteri, *kool*-ta-ree. After Dr
  Thomas Coulter.

Romulea, rom-*ew*-lee-a. *Iridaceae.*
After Romulus, founder of Rome.
Semi-hardy cormous herbs.
  bulbocodium, bul-bo-*ko*-dee-um.
  With woolly bulbs.
  clusiana, kloo-zee-*a*-na. After
  Clusius.
  flava, *fla*-va. Yellow.
  requienii, rek-wee-*en*-ee-ee. After
  Requien.
  rosea, ro-see-a. Rose-coloured
  (flowers).

Rosa, *ro*-sa. *Rosaceae.* L. name.
Deciduous or semi-evergreen shrubs
and climbers. Rose.
  x *alba,* *al*-ba. White. White Rose.
  banksiae, *banks*-ee-ee. After Lady
  Banks, wife of Sir Joseph. Banksian
  Rose.
  brunonii, broo-*non*-ee-ee. After
  Robert Brown. Himalayan Musk
  Rose.

Rosa canina

  canina, ka-*neen*-a. Of dogs. Dog
  Rose, Common Brier.
  centifolia, sent-i-*fo*-lee-a. With a
  hundred leaves (petals). Provence
  Rose, Cabbage Rose.
  chinensis, chin-*en*-sis. Of China.
  damascena, da-ma-*see*-na. Of
  Damascus. Summer Damask Rose.
  ecae, *ee*-see. After Mrs E. C.
  Aitchinson (E.C.A.).
  eglanteria, eg-lan-*te*-ree-a. Prickly
  Eglantine.
  elegantula, e-le-*gant*-ew-la. Elegant.
  filipes, *fi*-li-pees. Slender-stalked.
  foetida, *fee*-ti-da. Foetid. Austrian
  Brier.
  gallica, *gal*-i-ka. Of France. French
  Rose, Red Rose.
  glauca, *glow*-ka. Glaucous.
  x *harisonii,* ha-ri-*son*-ee-ee. After
  George Folliot Harison.
  helenae, *he*-len-ee. After Mrs Ernest
  Wilson, Ellen.
  moyesii, *moyz*-ee-ee. After the Rev.
  J. Moyes.
  nitida, *ni*-ti-da. Shining (leaves).
  x *odorata,* o-do-*ra*-ta. Scented.
  omeiensis, o-mee-*en*-sis. Of the

Omei Shan, China
palustris, pa-*lus*-tris. Growing in
bogs. Swamp Rose.
pimpinellifolia, pim-pi-nel-i-*fo*-lee-a.
Pimpinella-leaved. Burnet Rose,
Scotch Rose.
rubiginosa, roo-bi-ji-*no*-sa. Rusty.
rubrifolia, ru-bri-*fo*-lee-a. Red-
leaved.
rugosa, roo-*go*-sa. Wrinkled
(leaves). Japanese Rose.
sericea, se-*ri*-see-a. Silky-hairy.
spinosissima, spin-o-*sis*-i-ma. Very
spiny.
wichuraiana, wi-kewr-ra-ee-*a*-na.
After Max Wichura. Memorial Rose.
xanthina, zan-*theen*-a. Yellow (flowers).

Roscoea, ros-*ko*-ee-a. *Zingiberaceae.*
After William Roscoe. Semi-hardy,
tuberous perennial herbs.
auriculata, aw-rik-ew-*la*-ta. With
auricles.
cautleyoides, kawt-lee-*oi*-deez.
Cautleya-like.
humeana, hew-mee-*a*-na. After
David Hume.
purpurea, pur-*pur*-ree-a. Purple
(flowers).

Rosmarinus, ros-ma-*reen*-us.
Labiatae. From L. *ros* (dew) and *mari-
nus* (of the sea). Evergreen shrubs.
officinalis, o-fi-si-*na*-lis. Sold in
shops. Rosemary.

Rubus, *rub*-us. *Rosaceae.* L. for
blackberry. Deciduous, semi-evergreen
shrubs and climbers.
arcticus, ark-*tik*-us. Of the Polar
regions. Crimson Bramble.
caesius, see-zee-us. Bluish-grey.
Dewberry.
calycinoides, ka-li-si-*noi*-deez. Like
R. calycinus.

Rubus idaeus

cockburnianus, ko-burn-ee-*a*-nus.
After Cockburn.
deliciosus, de-li-see-*o*-sus.
Delightful. Rocky Mountain
Raspberry.
idaeus, ie-*de*-us. Of Mt. Ida.
Raspberry.
loganobaccus, lo-ga-no-*ba*-kus.
After James Harvey Logan and L.
baccus (berry). Loganberry.
odoratus, o-do-*ra*-tus. Scented.
Flowering Raspberry, Thimbleberry.
thibetanus, ti-bet-*a*-nus. Of Tibet.
tricolor, *tri*-ko-lor. Three-coloured.
ulmifolius, ul-mi-*fo*-lee-us. *Ulmus*-
leaved. Bramble.

Rudbeckia, rud-*bek*-ee-a. *Compositae.*
After Olof Rudbeck. Annual, biennial
and perennial herbs. Coneflower.
californica, ka-li-*forn*-i-ka. Of
California.
fulgida, *ful*-ji-da. Shining.
hirta, *hir*-ta. Hairy. Black-eyed
Susan.
laciniata, la-sin-ee-*a*-ta. Deeply cut
(leaves).
maxima, *max*-i-ma. Larger.

purpurea, pur-*pur*-ree-a. Purple.
subtomentosa, sub-to-men-*to*-sa.
Somewhat hairy. Sweet Coneflower.
triloba, tri-*lo*-ba. Three-lobed.
Brown-eyed Susan.

Ruellia, roo-*el*-ee-a. *Acanthaceae.*
After Jean Ruel. Tender, perennial
herbs and evergreen sub-shrubs.
   amoena, a-*mee*-na. Pleasant.
   devosiana, de-vos-ee-*a*-na. After
   Cornelius de Vos.
   macrantha, ma-*kranth*-a. Large-
   flowered. Christmas Pride.
   makoyana, ma-koy-*a*-na. After Jacob
   Makoy. Trailing Velvet Plant.
   portellae, por-*tel*-ee. After Francisco
   Portella.

Rumex, *ru*-mex. *Polygonaceae.* L.
name for *R. acetosa.* Biennial and
perennial herbs. Dock, Sorrel.
   alpinus, al-*pie*-nus. Alpine. Monk's
   rhubarb.
   acetosa, a-see-*to*-sa. Old name for
   plants with acid leaves. Garden
   Sorrel.
   scutatus, skoo-*ta*-tus. Shield-bearing.
   French sorrel.

Ruscus, *rus*-kus. *Liliaceae.* L. name.
Evergreen shrubs.
   aculeatus, a-kew-lee-*a*-tus. Prickly.
   Butcher's Broom, Box Holly.
   hypoglossum, hi-po-*glos*-um.

Rumex acetosa

Beneath the tongue (flowers are
under the tongue-like bract).

Russelia, rus-*el*-ee-a.
Scrophulariaceae. After Dr Alexander
Russel. Tender, evergreen shrubs and
sub-shrubs.
   equisetiformis, e-kwi-see-ti-*form*-is.
   Equisetum-like. Coral Plant,
   Firecracker Plant.
   lilacina, li-la-*seen*-a. Lilac.

Ruta, *roo*-ta. *Rutaceae.* L. name.
Evergreen sub-shrub.
   graveolens, gra-*vee*-o-lenz. Strong-
   smelling. Rue.

S

Sagina, sa-*jeen*-a. *Caryophyllaceae*. From L. *sagina* (fodder). Annual and evergreen perennial herbs. Pearlwort.
boydii, boyd-ee-ee. After William Brack Boyd.
procumbens, pro-*kum*-benz. Prostrate.
subulata, sub-ew-*la*-ta. Awl-shaped (leaves).

Sagittaria, sa-ji-*ta*-ree-a. *Alismataceae*. From L. *sagitta* (arrow), after the leaf shape. Deciduous and perennial aquatic herbs. Arrowhead.
japonica, ja-*pon*-i-ka. Of Japan.
latifolia, la-ti-*fo*-lee-a. Broad-leaved. Duck Potato.
sagittifolia, sa-gi-ti-*fo*-lee-a. With arrow-shaped leaves. Old World Arrowhead.

Sagittaria sagittifolia

Saintpaulia, saynt-*pawl*-ee-a. *Gesneriaceae*. After Baron von Saint Paul-Illaire. Tender, evergreen perennial herbs. African Violet.
ionantha, ie-on-*anth*-a. With violet flowers. African Violet.

Salix, sa-lix. *Salicaceae*. L. name. Deciduous trees and shrubs. Willow.
acutifolia, a-kew-ti-*fo*-lee-a. With pointed leaves.
aegyptiaca, ee-jip-tee-*a*-ka. Of Egypt. Musk Willow.
alba, al-ba. White (leaves). White Willow.

Salix alba

arctica, ark-*tik*-a. Of the Polar regions. Arctic Willow.
babylonica, bab-ill-*on*-ik-a. Of Babylon. Babylon Weeping Willow.
boydii, boyd-ee-ee. After William Brack Boyd.
caesia, see-zee-a. Bluish-grey. Blue Willow.
candida, kan-did-a. Shining. Sage Willow.

caprea, kap-ree-a. Of goats. Goat
Willow, Pussy Willow.
daphnoides, daf-*noi*-deez. Laurel-
like. Violet Willow.
fragilis, fra-ji-lis. Fragile (shoots).
Crack Willow, Brittle Willow.
hastata, has-*ta*-ta. Spear-shaped
(leaves). Halberd Willow.
lanata, la-*na*-ta. Woolly. Woolly
Willow.
matsudana, mat-soo-*da*-na. After
Matsuda. Peking Willow.
nigra, nig-ra. Black. Black Willow.
pentandra, pen-*tan*-dra. With five
stamens. Bay Willow, Laurel
Willow.
purpurea, pur-*pur*-ree-a. Purple
(shoots). Purple Willow, Basket
Willow.
repens, ree-penz. Creeping.
Creeping Willow.
reticulata, re-tik-ew-*la*-ta. Net-
veined (leaves).
viminalis, vi-min-*a*-lis. With long,
slender shoots. Common Osier,
Hemp Willow.

Salpiglossis, sal-pi-*glos*-is.
Solanaceae. From Gk. *salpinx* (trumpet) and *glossa* (tongue). Tender annual and biennial herbs.
sinuata, sin-ew-*a*-ta. Wavy-edged
(leaves). Painted Tongue.

Salvia, *sal*-vee-a. *Labiatae.* From L.
salvus (safe), after its healing properties. Annual and perennial herbs and
semi-evergreen shrubs. Sage.
aethiopsis, ee-thi-*op*-sis. Of Africa.
African Sage.
argentea, ar-*jen*-tee-a. Silvery
(leaves).
azurea, a-*zew*-ree-a. Deep blue
(flowers).
grandiflora, grand-i-*flo*-ra. Large-
flowered.

Salvia pratensis

blepharophylla, blef-a-ro-*fil*-la. With
fringed leaves.
caerulea, see-*ru*-lee-a. Dark blue.
farinacea, fa-ree-*na*-see-a. Mealy.
Mealy Sage.
fulgens, ful-jenz. Shining (flowers).
greggii, greg-ee-ee. After Dr John
Gregg. Autumn Sage.
horminum, hor-*mie*-num. Gk. for sage.
involucrata, in-vo-loo-*kra*-ta. With
bracts around the flowers.
microphylla, mik-ro-*fil*-la. Small-
leaved.
nemorosa, ne-mo-*ro*-sa. Of woods.
officinalis, o-fi-si-*na*-lis. Sold in
shops. Common Sage.
patens, pa-tenz. Spreading (flowers).
pratensis, pra-*ten*-sis. Of meadows.
Meadow Clary.
rutilans, roo-ti-lanz. Reddish (flowers). Pineapple-scented Sage.
sclarea, skla-ree-a. Clear. Clary.
splendens, splen-denz. Splendid.
Scarlet Sage.
uliginosa, ew-li-gi-*no*-sa. Of marshes.

Salvinia, sal-*veen*-ee-a. *Salviniaceae.*
After Professor Antonio Salvini.

Tender, deciduous floating ferns.
auriculata, aw-rik-ew-*la*-ta.
Auricled.
natans, na-tanz. Floating.

Sambucus, sam-*bew*-kus.
Caprifoliaceae. L. for elder.
Deciduous trees and shrubs. Elder,
Elderberry.
caerulea, see-*ru*-lee-a. Dark Blue.
Blue Elder.
canadensis, kan-a-*den*-sis. Of
Canada. American Elder, Sweet
Elder.
nigra, nig-ra. Black (fruits).
Common Elder, Elderberry.
racemosa, ra-see-*mo*-sa. In racemes.
Red-berried Elder.

Sambucus nigra

Sandersonia, san-der-*son*-ee-a.
Liliaceae. After John Sanderson.
Semi-hardy, deciduous climber.
aurantiaca, aw-ran-tee-a-ka. Orange
(leaves). Chinese Lanterns.

Sanguinaria, sang-gwi-*na*-ree-a.
Papaveraceae. From L. *sanguis*
(blood), after the red sap. Perennial

rhizomatous herb.
canadensis, kan-a-*den*-sis. Of
Canada. Bloodroot.

Sanguisorba, sang-gwi-*sor*-ba.
Rosaceae. From L. *sanguis* (blood)
and *sorbeo* (absorb). Perennial herbs.
Burnet.
canadensis, kan-a-*den*-sis. Of
Canada. Canadian Burnet.
obtusa, ob-*tew*-sa. Blunt (leaflets).
officinalis, o-fi-si-*na*-lis. Sold in
shops. Great Burnet.

Sansevieria, san-sev-ee-*e*-ree-a.
Agavaceae. After Raimond de
Sansgrio, Prince of Sansevier. Tender,
evergreen rhizomatous herbs.
cylindrica, si-*lin*-dri-ka. Cylindrical.
trifasciata, tri-fas-ee-a-ta. In three
bundles (flower clusters). Mother-in-
law's Tongue.

Santolina, san-to-*leen*-a. Compositae.
From L. *sanctum linum* (holy flax).
Evergreen shrubs.
chamaecyparissus, ka-mee-kew-pa-
ris-us. Dwarf cypress. Lavender
Cotton.
elegans, e-le-ganz. Elegant.
neopolitana, nee-a-po-li-*ta*-na. Of
Naples.
pinnata, pin-a-ta. Pinnate.
rosmarinifolia, ros-ma-reen-i-*fo*-lee-
a. *Rosmarinus*-leaved. Holy Flax.

Sanvitalia, san-vi-*ta*-lee-a.
Compositae. After Frederico Sanvitali.
Annual and perennial herbs.
procumbens, pro-*kum*-benz.
Prostrate. Creeping Zinnia.

Saponaria, sa-po-*na*-ree-a.
Caryophyllaceae. From L. *sapo*
(soap). Annual and perennial herbs.
Soapwort.

caespitosa, see-spi-*to*-sa. Tufted.
calabrica, ka-*la*-bri-ka. Of Calabria.
ocymoides, o-kim-*oi*-dees. *Ocimum*-like.
officinalis, o-fi-si-*na*-lis. Sold in shops. Soapwort.

Saponaria officinalis

Sarcococca, sar-ko-*ko*-ka. *Buxaceae.* From Gk. *sarcos* (flesh) and *kokkos* (berry). Sweet Box.
   confusa, kon-*few*-sa. Confused.
   digyna, di-*ji*-na. With two styles.
   hookeriana, hu-ka-ree-*a*-na. After Sir Joseph Hooker.
   humilis, hu-mi-lis. Low-growing.
   ruscifolia, rus-ki-*fo*-lee-a. *Ruscus*-leaved.
   saligna, sa-*lig*-na. Willow-like.

Sarracenia, sa-ra-*sen*-ee-a. *Sarraceniaceae.* After Michael Sarrasin. Insectivorous perennial herbs. Pitcher Plant.
   flava, fla-va. Yellow. Yellow Trumpet.
   minor, mi-nor. Smaller.
   purpurea, pur-*pur*-ree-a. Purple (pitchers). Common Pitcher Plant.

Sasa, *sa*-sa. *Gramineae.* Japanese name. Bamboos.
   albomarginata, al-bo-mar-ji-*na*-ta. White-margined.
   palmata, pal-*ma*-ta. Lobed like a hand (leaves).
   veitchii, veech-ee-ee. After Messrs Veitch.

Sassafras, *sas*-a-fras. *Lauraceae.* Deciduous trees.
   albidum, al-bi-dum. Whitish (under the leaves). Sassafras.

Satureja, sat-ew-*ree*-ee-a. *Labiatae.* L. name. Annual herbs and semi-evergreen sub-shrubs. Savory.
   hortensis, hor-*ten*-sis. Of gardens. Summer Savoury.
   montana, mon-*ta*-na. Of mountains. Winter Savoury.

Sauromatum, saw-*ro*-ma-tum. *Araceae.* From Gk. *sauros* (lizard). Tender, tuberous perennial herbs.
   guttatum, gu-*ta*-tum. Spotted (spathe).
   venosum, vee-*no*-sum. Conspicuously veined. Monarch of the East.

Saxegothaea, sax-ee-goth-*ee*-a. *Podocarpaceae.* After Prince Albert. Evergreen conifer.
   conspicua, con-*spik*-ew-a. Conspicuous. Prince Albert's Yew.

Saxifraga, sax-*if*-ra-ga. *Saxifragaceae.* From L. *saxum* (rock) and *frango* (break). Perennial herbs. Saxifrage.
   aizoides, ie-*zo*-i-deez. Like *Aizoon.*
   aspera, a-*spe*-ra. Rough. Rough Saxifrage.
   biflora, bi-*flo*-ra. Two-flowered.
   boydii, boyd-ee-ee. After William

Saxifraga aizoides

Brack Boyd.
brunonis, broo-*no*-nis. After Robert
Brown.
burseriana, bur-sa-ree-*a*-na. After
Joachim Burser.
callosa, ka-*lo*-sa. Calloused (leaves).
Limestone Saxifrage.
cochlearis, kok-lee-*a*-ris. Spoon-
shaped (leaves).
cortusifolia, kor-tew-si-*fo*-lee-a.
Cortusa-leaved.
fortunei, for-*tewn*-ee-ee. After
Robert Fortune.
granulata, gran-ew-*la*-ta. Composed
of minute grains.
grisebachii, gree-za-*bak*-ee-ee. After
Professor August Grisebach.
lingulata, ling-gew-*la*-ta. Tongue-like.
longifolia, long-i-*fo*-lee-a. Long-
leaved. Pyrenean Saxifrage.
moschata, mos-*ka*-ta. Musky.
oppositifolia, o-po-si-ti-*fo*-lee-a.
With opposite leaves. Purple
Saxifrage.
paniculata, pa-nik-ew-*la*-ta. With
flowers in panicles. Lifelong
Saxifrage.
sancta, sank-ta. Holy.
sarmentosa, sar-men-*to*-sa.

Producing runners.
stolonifera, sto-lo-*ni*-fe-ra. Bearing
stolons. Mother of Thousands,
Strawberry Geranium.
urbia, ur-bee-a. Of towns.

Scabiosa, skab-ee-*o*-sa. *Dipsacaceae.*
From L. *scabies* (itch), after the healing properties of the leaf. Annual and
perennial herbs. Scabious.
atropurpurea, at-ro-pur-*pur*-ree-a.
Deep purple. Sweet Scabious,
Pincushion Flower.
caucasica, kaw-*ka*-si-ka. Of the
Caucasus.
columbaria, kol-um-*ba*-re-a. Dove-
like.
graminifolia, gra-mi-ni-*fo*-lee-a.
With grass-like leaves.
lucida, loo-si-da. Shining.
ochroleuca, ok-ro-*loo*-ka. Yellowish-
white.
prolifera, pro-*li*-fe-ra. Proliferous.
Carmel Daisy.

Scabiosa columbaria

Schefflera, shef-*le*-ra. *Araliaceae.*
After J. C. Scheffler. Tender, evergreen
shrubs and trees.

actinophylla, ak-tin-o-*fil*-la. With rayed leaves. Umbrella Tree.
arboricola, ar-bo-*ri*-ko-la. Growing on trees.
digitata, di-gi-*ta*-ta. Lobed like a hand (leaves). Seven Fingers.

Schisandra, skis-*an*-dra. *Schisandraceae.* From Gk. *schizo* (divide) and *aner* (man). Deciduous climbers.
coccinea, kok-*kin*-ee-a. Scarlet. Wild Sarsparilla.
glaucescens, glow-*kes*-enz. Somewhat glaucous.
grandiflora, grand-i-*flo*-ra. Large-flowered.
propinqua, pro-*pin*-kwa. Related.
rubriflora, rub-ri-*flo*-ra. Red-flowered.
chinensis, chin-*en*-sis. Of China.

Schizanthus, skiz-*anth*-us. *Solanaceae.* From Gk. *schizo* (divide) and *anthos* (flower), after the divided corolla. Annual herbs. Poor Man's Orchid, Butterfly Flower.
hookeri, *huk*-a-ree. After W. J. Hooker.
pinnatus, pi-*na*-tus. Pinnate (leaves).

Schizophragma, ski-zo-*frag*-ma. *Hydrangeaceae.* From Gk. *schizo* (divide) and *phragma* (screen). Deciduous climbers.
hydrangeoides, hi-dran-jee-*oi*-deez. *Hydrangea*-like.
integrifolium, in-teg-ri-*fo*-lee-um. Entire-leaved.

Schizostylis, ski-zo-*sti*-lis. *Iridaceae.* From Gk. *schizo* (divide) and *stylis* (style), after the divided style. Rhizomatous perennial herb.
coccinea, kok-*kin*-ee-a. Scarlet. Kaffir Lily.

Schlumbergera, shlum-*ber*-ga-ra. *Cactaceae.* After Frederick Schlumberger.
bridgesii, bri-*jez*-ee-ee. After Thomas Bridges.
x *buckleyi,* *buk*-lee-ee. After W. Buckley. Christmas Cactus.
truncata, trun-*ka*-ta. Abruptly cut off.

Sciadopitys, skee-a-*do*-pi-tis. *Taxodiaceae.* From Gk. *skiados* (umbel) and *pitys* (fir tree). The leaves resemble an umbrella. Umbrella Pine, Japanese Umbrella Pine.
verticillata, ver-ti-si-*la*-ta. Whorled.

Scilla, *sil*-la. *Liliaceae.* From Gk. *Urginea maritima* (sea squill). Bulbous herbs.
bifolia, bi-*fo*-lee-a. Two-leaved.
campanulata, cam-pan-ew-*la*-ta. Bell-shaped.
chinensis, chin-*en*-sis. Of China.
hispanica, hi-*spa*-ni-ka. Of Spain.
mischtschenkoana, mi-cheng-ko-*a*-na. After Miczenko.
monophyllos, mo-no-*fil*-los. One-leaved.
natalensis, nat-al-*en*-sis. From Natal, S. Africa.
peruviana, pe-roo-vee-*a*-na. Of Peru.
sibirica, si-*bi*-ri-ka. Siberian.
violacea, vee-o-*la*-see-a. Violet.

Scindapsus, skin-*dap*-sus. *Araceae.* Gk. name for an ivy-like plant. Tender, evergreen climber.
argyraeus, ar-ji-*ree*-us. Silvery (leaves).
pictus, *pik*-tus. Painted (leaves). Silver Vine.

Scrophularia, skro-few-*la*-ree-a. *Scrophulariaceae.* From L. *scrofula* (wart), after its alleged healing properties. Perennial herbs and sub-shrubs.

auriculata, aw-rik-ew-*la*-ta. Auricled (leaves). Water Figwort.

Scrophularia auriculata

Scutellaria, sku-te-*la*-ree-a, *Labiatae.* From L. *scutella* (small dish). Tender and hardy rhizomatous perennial herbs. Skullcap.

   *indica, in-*di-ka. Of India.
   orientalis, o-ree-en-*ta*-lis. Eastern.
   scordiifolia, skor-dee-i-*fo*-lee-a. With *Scordium*-like leaves.

Sedum, *se*-dum, *Crassulaceae.* Classical name for several succulent plants from L. *sedo* (sit). Tender and hardy succulents, annuals and ever-green biennials..

Sedum acre

acre, ak-ree. Sharp-tasting. Stonecrop.
album, al-bum. White.
bellum, bel-um. Beautiful.
brevifolium, bre-vi-*fo*-lee-um. Short-leaved.
caeruleum, see-*ru*-lee-um. Dark blue.
cauticolum, kaw-*ti*-ko-lum. Growing on cliffs.
dasyphyllum, das-i-*fil*-lum. With hairy leaves.
dendroideum, den-*droi*-dee-um. Tree-like.
ewersii, ew-*werz*-ee-ee. After Joseph Ewers .
floriferum, flo-*ri*-fe-rum. Floriferous.
kamtschaticum, kamt-*sha*-ti-kum. Of Kamtchatka.
lineare, li-nee-*a*-ree. Linear (leaves).
morganianum, mor-gan-ee-*a*-num. After Dr Meredith Morgan.
oreganum, o-ree-*ga*-num. Of Oregon.
pachyphyllum, pa-ki-*fil*-lum. Thick-leaved.
populifolium, po-pew-li-*fo*-lee-um. *Populus*-leaved.
praealtum, pree-*al*-tum. Very tall.
reflexum, re-*flex*-um. Reflexed (leaves).
rosea, ro-see-a. Rose-coloured
x *rubrotinctum,* rub-ro-*tink*-tum. Red-tinged (leaves).
rupestre, roo-*pes*-tree. Growing on rocks.
sieboldii, see-*bold*-ee-ee. After Siebold.
spathulifolium, spath-ew-li-*fo*-lee-um. With spatula-shaped leaves.
spectabile, spek-*ta*-bi-lee. Spectacular.

Selaginella, se-la-ji-*nel*-a, *Selaginellaceae.* Diminutive of *selago.* Tender, evergreen moss-like perennials. Spike Moss.

apoda, a-*pod*-a. Stalkless. Basket Spike Moss.

kraussiana, krows-ee-*a*-na. After Ferdinand F. Krauss. Trailing Spike Moss.

lepidophylla, le-pi-do-*fil*-la. With scale-like leaves. Rose of Jericho.

Selenicereus, se-lee-nee-*see*-ree-us, *Cactaceae.* From Gk. *selene* (moon) and *Cereus.*

grandiflorus, grand-i-*flo*-rus. Large-flowered.

megalanthus, me-ga-*lanth*-us. Large-flowered.

pteranthus, te-*ranth*-us. With winged flowers.

Selinum, se-*leen*-um, *Umbelliferae.* From Gk. *selinon* (celery). Perennial herb.

tenuifolium, ten-ew-i-*fo*-lee-um. With finely divided leaves.

Sempervivum, sem-per-*veev*-um, *Crassulaceae.* From L. *semper* (always) and *vivus* (alive). Evergreen perennials. Houseleek.

Sempervivum tectorum

arachnoideum, a-rak-*noi*-dee-um. With hairs like a spiders-web.

ciliosum, sil-ee-*o*-sum. Slightly fringed.

grandiflorum, grand-i-*flo*-rum. Large-flowered.

montanum, mon-*ta*-num. Of mountains.

tectorum, tek-*to*-rum. Growing on roofs. Common Houseleek.

Senecio, se-*ne*-see-o, *Compositae.* From L. *senex* (old man), after the white seed heads. Herbs, tender succulents and evergreen shrubs.

articulatus, ar-tik-ew-*la*-tus. Jointed. Candle Plant.

bicolor, *bi*-ko-lor. Two-coloured.

cineraria, si-ne-*ra*-ee-a. Ash-coloured (leaves).

clivorum, klie-*vor*-um. Of the hills.

confusus, kon-*few*-sus. Confused

compactus, com-*pak*-tus. Compact.

cruentus, kroo-*en*-tus. Blood-red (flowers).

doronicum, do-*ro*-ni-kum. From *Doronicum.*

elegans, *e*-le-ganz. Elegant.

fulgens, *ful*-jenz. Shining.

grandifolius, gran-di-*fo*-lee-us. With large leaves.

haworthii, hay-*werth*-ee-ee. After Haworth. Cocoon Plant.

x *hybridus,* *hib*-ri-dus. Hybrid. Cineraria.

laxifolius, lax-i-*fo*-lee-us. Loose-leaved. New Zealand.

macroglossus, mak-ro-*glos*-us. Large-tongued. Wax Vine.

mikanioides, mi-ka-nee-*oi*-deez. *Mikania*-like.

monroi, mon-*ro*-ee. After Sir David Monro.

pendulus, *pen*-dew-lus. Pendulous.

pulcher, *pul*-ker. Pretty.

scandens, *skan*-denz. Climbing.

serpens, ser-penz. Snake-like. Blue chalksticks.
tanguticus, tan-*gew*-ti-kus. Of Gansu, China.

Sequoia, se-*kwoy*-a. *Taxodiaceae.* After Sequoiah, a Cherokee Indian name. Evergreen conifer.
sempervirens, sem-per-*vi*-renz. Evergreen. California Redwood.

Sequoiadendron, se-kwoy-a-*den*-dron. *Taxodiaceae.* From *Sequoia* and Gk. *dendron* (tree). Evergreen conifer. Giant Sequoia.
giganteum, ji-*gan*-tee-um. Very large.

Shibataea, shi-ba-*tee*-a, *Gramineae.* After Keita Shibata. Bamboo.
kumasasa, kew-ma-*sa*-sa. Japanese name.

Shortia, *short*-ee-a. *Diapensiaceae.* After Charles W. Short. Evergreen, perennial herbs.
galacifolia, ga-las-i-*fo*-lee-a. With *Galax*-like leaves. Oconee Bells.
ilicifolia, ee-li-si-*fo*-lee-a. *Ilex*-leaved.
soldanelloides, sol-da-nel-*oi*-deez. *Soldanella*-like. Fringe Bell.
uniflora, ew-ni-*flo*-ra. With one flower. Nippon Bells.

Sidalcea, see-*dal*-see-a. *Malvaceae.* From *Sida* and *Alcea.* Perennial herbs. Prairie Mallow.
campestre, kam-*pes*-tree. Of fields. Meadow Sidalcea.
candida, kan-di-da. White (flowers). White Prairie Mallow.
malviflora, mal-vi-*flo*-ra. *Malva*-flowered. Checkerbloom.

Silene, si-*lee*-nee. *Caryophyllaceae.* Gk. name for another plant. Annual

and perennial herbs. Campion, Catchfly.
acaulis, a-*kaw*-lis. Stemless. Moss Campion.
alpestris, al-*pes*-tris. Of lower mountains.
armeria, ar-*me*-ree-a. Growing near the sea.
coeli-rosea, see-lee-*ro*-see-a. Rose of Heaven.
compacta, com-*pak*-ta. Compact.
dioica, dee-o-*ee*-ka. Dioecious. Red Campion.
hookeri, huk-a-ree. After W. J. Hooker.
maritima, ma-*ri*-ti-ma. Growing near the sea.
pendula, pen-dew-la. Pendulous (flowers).
vulgaris, vul-*ga*-ris. Common.

Silene acaulis

Silybum, *si*-li-bum. *Compositae.* From Gk. *silybon.* Annual or biennial herb.
marianum, ma-ree-*a*-num. Of the Virgin Mary. Blessed Thistle. Our Lady's Milk Thistle.

Sinningia, si-*nin*-gee-a. *Gesneriaceae.* After William Sinning. Tender, perennial herbs and deciduous sub-shrubs.
barbata, bar-*ba*-ta. Bearded.
cardinalis, kar-di-*na*-lis. Scarlet.
concinna, kon-*sin*-a. Elegant.
eumorpha, ew-*morf*-a. Of good shape.

macropoda, ma-*kro*-po-da.With a large stalk.
pusilla, pu-*sil*-la. Dwarf.
speciosa, spes-ee-*o*-sa. Showy. Gloxinia.

Sisyrinchium, si-si-*rin*-kee-um. *Iridaceae.* Gk. name for another plant. Annual and perennial herbs.
angustifolium, ang-gus-ti-*fo*-lee-um. Narrow-leaved. Blue-eyed Grass.
bellum, be-lum. Beautiful. Californian Blue-eyed Grass.
bermudiana, ber-mew-dee-*a*-na. Of Bermuda.
brachypus, bra-ki-pus. Short-stalked (flowers).
californicum, ka-li-*forn*-i-kum. Of California.
douglasii, dug-*las*-ee-ee. After Douglas.
grandiflora, gran-di-*flo*-ra. With large flowers.
odoratissimum, o-do-ra-*tis*-i-mum. Highly scented.
striatum, stri-*a*-tum. Striped (flowers).

Sisyrinchium angustifolium

Skimmia, *skim*-ee-a. *Rutaceae.* From the Japanese *Miyami-Shikimi.* Evergreen trees and shrubs.
anquetilia, an-kwe-*ti*-lee-a. After Anquetil-Duperron.
x *foremanii,* for-*man*-ee-ee. After Foreman.
japonica, ja-*pon*-i-ka. Of Japan.
reevesiana, reev-zee-*a*-na. After John Reeves.

Smilacina, smee-la-*seen*-a. *Liliaceae.* Diminutive of *Smilax.* Perennial herbs. False Solomon's Seal.
racemosa, ra-see-*mo*-sa. With flowers in racemes. False Spikenard, Treacleberry.
stellata, ste-*la*-ta. Star-like. Star-flowered Lily-of-the-Valley.

Smilax, *smi*-lax. *Liliaceae.* The Gk. name. Evergreen, deciduous and herbaceous climbers.
aspera, a-*spe*-ra. Rough (stems).
excelsa, ex-*sel*-sa. Tall.
rotundifolia, ro-tun-di-*fo*-lee-a. Round-leaved.

Smithiantha, smith-ee-*anth*-a. *Gesneriaceae.* After Matilda Smith. Tender, rhizomatous perennial herbs. Temple Bells.
cinnabarina, si-na-ba-*reen*-a. Scarlet.
zebrina, ze-*breen*-a. Striped (leaves).

Solanum, so-*la*-num. *Solanaceae.* L. name. Annual and perennial herbs, shrubs and climbers. Nightshade.
capsicastrum, kap-si-*kas*-trum. Pepper-like (fruit). False Jerusalem Cherry.
crispum, kris-pum. Wavy-edged (leaves).
jasminoides, jas-min-*oi*-dees. Jasmine-like. Potato Vine.

melongena, me-lon-*jee*-na. From old French *melongene.* Aubergine, Egg Plant.

pseudocapsicum, soo-do-*kap*-si-kum. False *Capsicum.* Winter Cherry, Jerusalem Cherry.

tuberosum, tew-be-*ro*-sum. Tuberous. Potato.

Soldanella, sol-da-*nel*-la. *Primulaceae.* From Italian *soldo* (small coin), after the rounded leaves. Evergreen perennial herbs.

alpina, al-*pie*-na. Alpine.

minima, mi-ni-ma. Smaller.

montana, mon-*ta*-na. Of mountains.

pusilla, pu-*sil*-la. Dwarf.

villosa, vi-*lo*-sa. Softly hairy.

Soleirolia, so-lee-*rol*-ee-a. *Urticaceae.* After Joseph Francois Soleirol. Evergreen, perennial herb. Baby's Tears, Mind your own Business.

soleirolii, so-lee-*rol*-ee-ee. As above.

Solidago, so-li-*da*-go. *Compositae.* From L. *solido* (strengthen), after its healing properties. Perennial herbs. Golden Rod.

bicolor, bi-ko-lor. Two-coloured. Silver Rod.

canadensis, kan-a-*den*-sis. Of Canada.

virgaurea, virg-*aw*-ree-a. A golden rod.

Sollya, *so*-lee-a. *Pittosporaceae.* After Richard Horsman Solly. Tender, evergreen climbers.

heterophylla, he-te-ro-*fil*-la.With variable leaves. Bluebell Creeper.

parviflora, par-vi-*flo*-ra. Small-flowered.

Sonerila, so-*ne*-ri-la. *Melastomataceae.* From *soneri-ila,* the Malabar name. Tender, evergreen perennials and shrubs.

grandiflora, gran-di-*flo*-ra. With large flowers.

margaritacea, mar-ga-ri-*ta*-see-a. Pearly (leaves).

Sophora, so-*fo*-ra. *Leguminosae.* From the Arabic name. Deciduous and evergreen trees and shrubs.

davidii, da-*vid*-ee-ee. After David.

japonica, ja-*pon*-i-ka. Of Japan. Japanese Pagoda Tree.

microphylla, mik-ro-*fil*-la. Small-leaved.

tetraptera, tet-*rap*-te-ra. Four-winged (pod). Kowhai.

Sophronitis, so-*fro*-ni-tis. *Orchidaceae.* From Gk. *sophron* (modest). Greenhouse orchids.

cernua, sern-ew-a. Nodding.

coccinea, kok-*kin*-ee-a. Scarlet.

Sorbaria, sor-*ba*-ree-a. *Rosaceae.* From L. *Sorbus.* Deciduous shrubs. False Spiraea.

Solidago virgaurea

aitchisonii, aych-i-*son*-ee-ee. After Dr John Aitchison.
arborea, ar-*bo*-ree-a. Tree-like.
grandiflora, gran-di-*flo*-ra. With large flowers.
sorbifolia, sor-bi-*fo*-lee-a. *Sorbus*-leaved.

Sorbus, *sor*-bus. *Rosaceae.* L. name for *S. domestica.* Deciduous trees and shrubs. Mountain Ash.
alnifolia, al-ni-*fo*-lee-a. *Alnus*-leaved.
americana, a-me-ri-*ka*-na. Of America. American Mountain Ash.
aucuparia, aw-kew-*pa*-ree-a. Bird-catching. Common Mountain Ash.
cashmiriana, kash-mi-ree-*a*-na. Of Kashmir.
commixta, kom-*mix*-ta. Mixed together.
cuspidata, kus-pi-*da*-ta. Abruptly sharp-pointed (leaves).
decora, de-*ko*-ra. Beautiful.
discolor, dis-ko-lor. Of two colours.
domestica, do-*mes*-ti-ka. Cultivated.
esserteauiana, es-er-toe-ee-*a*-na. After Dr Esserteau.
hupehensis, hew-pee-*hen*-sis. Of Hupeh, China.
insignis, in-*sig*-nis. Notable.

Sorbus latifolia

latifolia, la-ti-*fo*-lee-a. Broad-leaved.
poteriifolia, po-te-ree-i-*fo*-lee-a. With *Poterium*-like leaves.
reducta, re-*duk*-ta. Dwarf.
sargentiana, sar-jent-ee-*a*-na. After Sargent.
scalaris, ska-*la*-ris. Ladder-like (leaves).
x *thuringiaca,* thu-ring-gee-*a*-sa. Of Thuringia, Germany.
torminalis, tor-mi-*na*-lis. Effective against colic. Chequer Tree.
vilmorinii, vil-mo-*rin*-ee-ee. After Maurice Vilmorin.

Sparaxis, spa-*rax*-is. *Iridaceae.* From Gk. *sparasso* (tear). Cormous, semi-hardy perennial herbs. Harlequin Flower.
elegans, e-le-ganz. Elegant.
grandiflora, grand-i-*flo*-ra. Large-flowered.

Sparmannia, spar-*man*-ee-a. *Tiliaceae.* After Dr Andreas Sparrman. Tender evergreen trees and shrubs.
africana, af-ri-*ka*-na. African. African Hemp.

Spartina, spar-*teen*-a. *Gramineae.* From Gk. *spartion* (esparto grass). Perennial grass.
pectinata, pek-ti-*na*-ta. Comb-like.

Spartium, *spar*-tee-um. *Leguminosae.* From Gk. *spartion* (esparto grass). Deciduous, almost leafless shrub.
junceum, jun-see-um. Rush-like. Spanish Broom.

Spathiphyllum, spa-thi-*fil*-lum. *Araceae.* From Gk. *spathe* and *phyllon* (leaf). Tender herb.
blandum, blan-dum. Pleasant.
floribundum, flo-ri-*bun*-dum. Profusely flowering.

wallisii, wol-*is*-ee-ee. After Gustave Wallis.

Sphaeralcea, sfee-*ral*-see-a. *Malvaceae.* From Gk. *sphaira* (globe) and *Alcea,* after the spherical fruit. Semi-hardy perennial herbs and sub-shrubs.

ambigua, am-*big*-ew-a. Doubtful.
coccinea, kok-*kin*-ee-a. Scarlet.
munroana, mun-ro-*a*-na. After Munro.

Spinacia, spee-*na*-see-a. *Chenopodiaceae.* From L. *spina* (spine), after the spiny husks. Annual and biennial herb.

oleracea, o-le-*ra*-see-a. Vegetable-like. Spinach.

Spiraea, spee-*ree*-a. *Rosaceae.* From Gk. *speiraira* (used for garlands). Deciduous or semi-evergreen shrubs. Spirea.

alba, *al*-ba. White. Meadowsweet.
arguta, ar-*gew*-ta. Sharply toothed (leaves).
bella, *be*-la. Pretty.

Spiraea salicifolia

canescens, ka-*nes*-enz. Greyish-white hairs.
densiflora, dens-i-*flo*-ra. Densely-flowered.
douglasii, dug-*las*-ee-ee. After Douglas.
japonica, ja-*pon*-i-ka. Of Japan.
nipponica, ni-*pon*-i-ka. Of Japan.
prunifolia, proo-ni-*fo*-lee-a. *Prunus-*leaved.
salicifolia, sa-li-si-*fo*-lee-a. *Salix-*leaved. Bridewort.
trilobata tri-lo-*ba*-ta. Three-lobed (leaves).

Stachys, sta-kis. *Labiatae.* From Gk. *stachys* (spike). Tender to hardy perennial herbs and sub-shrubs.

affinis, a-*fee*-nis. Related to. Chinese Artichoke.
coccinea, kok-*kin*-ee-a. Scarlet.
byzantina, bi-zan-*teen*-a. Of Byzantium. Lamb's Tongue.
corsica, *kor*-si-ka. Of Corsica.
macrantha, ma-*kranth*-a. Large-flowered.
monieri, mo-nee-*e*-ree. After Monier.
officinalis, o-fi-si-*na*-lis. Sold in shops. Bishop's Wort.

Stachyurus, sta-kee-*ew*-rus. *Stachyuraceae.* From Gk. *stachys* (spike) and *oura* (tail). Deciduous shrubs.

chinensis, chin-*en*-sis. Of China.
praecox, *pree*-kox. Early (flowering).

Stanhopea, stan-*ho*-pee-a. *Orchidaceae.* After Philip Henry, 4th Earl of Stanhope. Greenhouse orchids.

grandiflora, grand-i-*flo*-ra. Large-flowered.
oculata, ok-ew-*la*-ta. With an eye.
tigrina, ti-*green*-a. Striped like a tiger.
wardii, *ward*-ee-ee. After Ward.

Stapelia, sta-*pel*-ee-a. *Asclepiadaceae.*
After Johannes von Stapel. Tender succulents.
 gigantea, ji-*gan*-tee-a. Very large.
 Giant Stapelia.
 grandiflora, grand-i-*flo*-ra. Large-
 flowered.
 variegata, va-ree-a-*ga*-ta. Variegated
 (corolla). Starfish Plant.

Staphylea, sta-*fi*-lee-a. *Staphyleaceae.*
From Gk. *staphyle* (cluster).
Deciduous trees and shrubs. Bladder
Nut.
 colchica, kol-ch-ka. Of Colchis.
 holocarpa, ho-lo-*kar*-pa. With an
 unlobed fruit.
 pinnata, pi-*na*-ta. Pinnate (leaves).
 Bladder Nut.

Stauntonia, stawn-*ton*-ee-a.
Lardizabalaceae. After Sir George
Leonard Staunton. Evergreen climber.
 hexaphylla, hex-a-*fil*-la. With six
 leaves.

Stenocarpus, sten-o-*kar*-pus.
Proteaceae. From Gk. *stenos* (narrow)
and *karpos* (fruit). Tender trees and
shrubs.
 sinuatus, sin-ew-*a*-tus. Wavy-edged
 (leaves). Australian Firewheel Tree.

Stephanandra, ste-fa-*nan*-dra.
Rosaceae. From Gk. *stephanos*
(crown) and *andros* (man), the stamens
form a wreath. Deciduous shrubs.
 incisa, in-*see*-sa. Deeply cut
 (leaves).
 tanakae, ta-*na*-kie. After Tanaka.

Stephanotis, ste-fa-*no*-tis.
Asclepiadaceae. From Gk. *stephanos*
(crown) and *otos* (ear). Tender, evergreen climbers.
 floribunda, flo-ri-*bun*-da. Profusely

flowering. Bridal Wreath,
Waxflower.

Sternbergia, stern-*berg*-ee-a.
Amaryllidaceae. After Count Kaspar
von Sternberg. Bulbous perennial
herbs. Autumn Daffodil.
 candida, kan-di-da. White.
 clusiana, klooz-ee-a-na. After
 Clusius.
 lutea, *loo*-tee-a. Yellow.
 sicula, sik-ew-la. Of Sicily.

Stewartia (Stuartia), stew-*art*-ee-a.
Theaceae. After John Stuart, 3rd Earl
of Bute. Deciduous trees and shrubs.
 malacodendron, mal-ak-o-*den*-dron.
 Silky Camellia.
 ovata, o-*va*-ta. Ovate (leaves).
 pseudocamellia, soo-do-ka-*mel*-ee-a.
 False *Camellia.* Japanese Stewartia.
 serrata, se-*ra*-ta. Saw-toothed
 (leaves).
 sinensis, si-*nen*-sis. Of China.

Stipa, *stee*-pa. *Gramineae.* From Gk.
tuppe (fibre), *S. tenacissima* is esparto
grass from which paper is made.
Perennial grasses. Needle Grass.
 barbata, bar-*ba*-ta. Bearded
 gigantea, ji-*gan*-tee-a. Very large.
 pennata, pe-*na*-ta. Feathery.
 European Feather Grass.
 pulcherrima, pul-*ke*-ri-ma. Very
 Pretty.
 splendens, splen-denz. Splendid.
 Chee Grass.

Stokesia, *stox*-ee-a. *Compositae.* After
Dr Jonathan Stokes. Perennial herb.
Stokes' Aster.
 laevis, *lee*-vis. Smooth.

Stratiotes, stra-tee-*o*-teez.
Hydrocharitaceae. From Gk. *stratiotes*
(soldier). Semi-evergreen, aquatic herb.

Stratiotes aloides

aloides, a-*lo*-i-deez. *Aloe*-like. Water Soldier.

Strelitzia, stre-*litz*-ee-a. *Strelitziaceae.* After Charlotte of Mecklenberg-Strelitz, Queen to George III. Tender herbaceous perennials.
  alba, al-ba. White (flowers).
  caudata, kaw-*da*-ta. With a tail.
  reginae, ree-*geen*-ee. Of the Queen. Bird of Paradise, Crane Flower.

Streptocarpus, strep-to-*kar*-pus. *Gesneriaceae.* From Gk. *streptos* (twisted) and *karpos* (fruit). The fruits are twisted. Tender perennial herbs. Cape Primrose.
  caulescens, kaw-*les*-enz. With a stem.
  dunnii, *dun*-ee-ee. After Edward Dunn.
  holstii, *holst*-ee-ee. After C. H. E. W. Holst.
  polyanthus, po-lee-*anth*-us. Many-flowered.
  saxorum, sax-*o*-rum. Growing on rocks.

Streptosolen, strep-to-*so*-len. *Solanaceae.* From Gk. *streptos* (twisted) and *solen* (tube). Tender, evergreen shrub.
  jamesonii, jaym-*son*-ee-ee. After Dr William Jameson. Marmalade Bush, Firebush.

Strobilanthes, stro-bi-*lanth*-eez. *Acanthaceae.* From Gk. *strobilos* (cone) and *anthos* (flower). Tender perennials and evergreen sub-shrubs.
  atropurpureus, a-tro-pur-*pur*-ree-us. Deep purple.
  dyerianus, die-a-ree-*a*-nus. After Sir William Thistleton-Dyer. Persian Shield.

Stromanthe, stro-*manth*-ee. *Marantaceae.* From Gk. *stroma* (bed) and *anthos* (flower). Tender and evergreen perennial herbs.
  sanguinea, sang-*gwin*-ee-a. Blood-red (bracts).

Stylophorum, sti-*lo*-fo-rum. *Papaveraceae.* From Gk. *stylos* (style) and *phoros* (bearing). Perennial herb.
  diphyllum, di-*fil*-lum. Two-leaved. Celandine Poppy, Wood Poppy.

Styrax, *sti*-rax. *Styracaceae.* Gk. name. Tender to hardy deciduous trees and shrubs.
  americanum, a-me-ri-*ka*-num. Of America.
  grandiflorum, gran-di-*flo*-rum. With large flowers.
  hemsleyana, hemz-lee-*a*-na. After William Botting Hemsley. China.
  japonicum, ja-*pon*-i-kum. Of Japan.
  obassia, o-*ba*-see-a. From the Japanese name.
  officinalis, o-fi-si-*na*-lis. Sold in shops.
  wilsonii, wil-*son*-ee-ee. After Wilson.

Sycopsis, si-*kop*-sis. *Hamamelidaceae.*
From Gk. *sykon* (fig) and *-opsis*
(resemblance). Evergreen trees and
shrubs.
 sinensis, si-*nen*-sis. Of China.

Symphoricarpos, sim-fo-ree-*kar*-pos.
Caprifoliaceae. From Gk. *symphorein*
(bear together) and *karpos* (fruit), after
the clustered fruits. Deciduous shrubs.
 albus, al-bus. White (fruit).
 Snowberry, Waxberry.
 x *chenaultii,* she-*nol*-tee-ee. After
 Chenault.
 mollis, mol-lis. Soft.
 occidentalis, ok-si-den-*ta*-lis.
 Western. Wolfberry.
 orbiculatus, or-bik-ew-*la*-tus.
 Orbicular (fruit). Coralberry, Indian
 Currant.

Symphytum, *sim*-fi-tum.
Boraginaceae. From Gk. *symphysis*
(growing together of bones) and *phyton* (plant), after its alleged healing
properties. Perennial herbs. Comfrey.
 caucasicum, kaw-*ka*-si-kum. Of the
 Caucasus.
 grandiflorum, grand-i-*flo*-rum.
 Large-flowered.
 x *uplandicum,* up-*land*-i-kum. Of
 Uppland, Sweden. Russian Comfrey.

Symplocos, *sim*-plo-kos.
Symplocaceae. From Gk. *symploke*
(connection). Evergreen or deciduous
trees and shrubs.
 paniculata, pa-nik-ew-*la*-ta. With
 flowers in panicles. Sapphire Berry.

Syngonium, sin-*gon*-ee-um. *Araceae.*
From Gk. *syn* (together) and *gone*
(womb). Tender, evergreen climbers.
 auritum, aw-*ree*-tum. Eared (the
 outer leaf segments). Five Fingers.
 Caribbean.
 hoffmannii, hoff-*man*-ee-ee. After
 Georg Franz Hoffman.
 podophyllum, po-do-*fil*-lum. With
 stoutly-stalked leaves. Arrowhead
 Vine.

Syringa, si-*ring*-ga. *Oleaceae.* From
Gk. *syrinx* (pipe) referring to the hollow stems. Deciduous trees and
shrubs. Lilac.
 x *chinensis,* chin-*en*-sis. Of China.
 Rouen Lilac.
 emodi, e-*mo*-dee. Of *Emodi Montes*
 (Himalayas).
 x *hyacinthiflora,* hi-a-sinth-i-*flo*-ra.
 With hyacinth-coloured flowers.
 josikaea, jo-si-*kee*-a. After Baroness
 von Josika. Hungarian Lilac.
 laciniata, la-sin-ee-*a*-ta. Deeply cut
 (leaves).
 meyeri, may-a-ree. After F. N. Meyer.
 microphylla, mik-ro-*fil*-la. Small-
 leaved.
 x *persica, per*-si-ka. Of Persia.
 reflexa, re-*flex*-a. Reflexed (corolla
 lobes).
 reticulatum, re-tik-ew-*la*-tum. Net-
 veined. Japanese Tree Lilac.
 velutina, vel-ew-*teen*-a. Velvety.
 vulgaris, vul-*ga*-ris. Common.
 Common Lilac.
 yunnanensis, yoo-nan-*en*-sis. Of
 Yunnan, China.

T

Tabebuia, ta-bee-*bew*-ee-a.
Bignoniaceae. From the Brazilian
name. Deciduous or evergreen trees.
  argentea, ar-*jen*-tee-a. Silvery. Tree
  of Gold.
  dubia, dub-ee-a. Doubtful.
  pallida, pa-li-da. Pale.
  rosea, ro-see-a. Rose-coloured. Pink
  Trumpet Tree.

Tacca, *ta*-ka. *Taccaceae*. From the
Malayan *taka*. Tender, rhizomatous
perennial herbs.
  chantrieri, shon-tree-*e*-ree. After
  Chantrier Frères. Bat Flower, Cat's
  Whiskers.
  leontopetaloides, lee-on-to-pe-ta-*loi*-
  deez. *Leontopetalon*-like. Indian
  Arrowroot.

Tagetes, ta-*gee*-teez. *Compositae*.
From the Etruscan *Tages*, grandson of
Jupiter. Annual herbs. Marigold.
  erecta, e-*rek*-ta. Erect. African
  Marigold, Big Marigold.
  lucida, loo-si-da. Bright. Sweet
  Mace.
  patula, pat-ew-la. Spreading. French
  Marigold.

Talinum, ta-*leen*-um. *Portulacaceae*.
Origin unknown. Tender, perennial
herbs. Fameflower.
  calycinum, ka-li-*see*-num. Calyx-
  shaped.
  guadalupense, gwa-da-loop-*en*-see.
  Of Guadaloupe.
  paniculatum, pa-nik-ew-*la*-tum.
  With flowers in panicles.
  Fameflower.
  reflexum, re-*flex*-um. Reflexed.

Tamarix, *ta*-ma-rix. *Tamaricaceae*. L.
name. Deciduous or evergreen trees
and shrubs. Tamarisk.
  gallica, ga-li-ka. Of France. Manna
  Plant.
  parviflora, par-vi-*flo*-ra. Small-flowered.
  ramosissima, ra-mo-*sis*-i-ma. Much
  branched.

Tamarix gallica

Tanacetum, tan-a-*set*-um.
Compositae. L. name. Perennial herbs.
  argenteum, ar-*jen*-tee-um. Silvery.
  coccineum, kok-*kin*-ee-um. Scarlet.
  Pyrethrum.
  corymbosum, ko-rim-*bo*-sum. With
  flowers in corymbs.
  densum, den-sum. Compact.
  haradjanii, ha-rad-*ya*-nee-ee. After
  Haradjian.
  macrophyllum, mak-ro-*fil*-um.
  Large-leaved.
  vulgare, vul-*ga*-ree. Common.
  Tansy.

Tanacetum vulgare

Tanakaea, tan-a-*kee*-a. *Saxifragaceae.*
After Yoshio Tanaka. Evergreen,
perennial herb.
  radicans, ra-di-kanz. With rooting
  stems.

Taxodium, tax-*o*-dee-um.
Taxodiaceae. From L. *Taxus* (Yew)
and Gk. *eidos* (resemblance).
Deciduous conifers. Swamp Cypress.

Taxus baccata

ascendens, a-*sen*-denz. Ascending.
Pond Cypress.
distichum, dis-ti-kum. In two ranks.
Swamp Cypress.

Taxus, *tax*-us. *Taxaceae.* L. and Gk.
name. Evergreen trees. Yew.
  baccata, ba-*ka*-ta. Berry-bearing. Yew.
  brevifolia, brev-i-*fo*-lee-a. With short
  leaves. Pacific Yew.
  cuspidata, kus-pi-*da*-ta. With a stiff
  point. Japanese Yew.

Tecoma, te-*ko*-ma. *Bignoniaceae.*
From the Mexican name. Tender, evergreen trees and shrubs.
  australis, aw-*stra*-lis. Southern.
  capensis, ka-*pen*-sis. Of the Cape of
  Good Hope.
  radicans, ra-di-kanz. With rooting
  stems.
  stans, stanz. Erect. Yellow Elder.

Tecophilaea, te-ko-fi-*lee*-a.
Tecophilaeaceae. After Tecophila
Billoti. Cormous perennial herbs.
  cyanocrocus, sie-an-o-*kro*-kus. Blue
  Crocus.

Tellima, *te*-li-ma. *Saxifragaceae.*
Anagram of *Mitella.* Semi-evergreen,
perennial herb.
  grandiflora, grand-i-*flo*-ra. Large-
  flowered.

Telopea, te-*lo*-pee-a. *Proteaceae.*
From Gk. *telopos* (viewed from afar).
Evergreen, tender and semi-hardy trees
and shrubs.
  speciosissima, spes-ee-o-*si*-si-ma.
  Very showy.
  truncata, trun-*ka*-ta. Abruptly cut off
  (leaves).

Tetrastigma, tet-ra-*stig*-ma. *Vitaceae.*
From Gk. *tetra* (four) and *stigma.*

Tender, evergreen climber.
voinierianum, vwan-ee-er-ee-*a*-num.
After M. Voinier. Lizard Plant.

Teucrium, *tewk*-ree-um. *Labiatae.*
After *Teucer,* King of Troy. Evergreen
or deciduous sub-shrubs and perennial
herbs. Germander, Wood Sage.
  aroanium, a-ro-*a*-nee-um. Of
  Aroania.
  canadense, kan-a-*den*-see. Of
  Canada. Wood Sage.
  chamaedrys, ka-*mee*-dris. *Chamaedrys*-like (dwarf oak). Wall
  Germander.
  fruticans, froo-ti-kanz. Shrubby.
  Tree Germander.
  marum, ma-rum. The Gk. name. Cat
  Thyme.
  polium, po-lee-um. Gk. name.

Teucrium chamaedrys

Thalictrum, tha-*lik*-trum.
Ranunculaceae. Gk. name for another
plant. Perennial herbs. Meadow Rue.
  alpinum, al-*pie*-num. Alpine.
  aquilegiifolium, a-kwi-lee-jee-i-*fo*-
  lee-um. *Aquilegia*-leaved.
  chelidonii, kel-i-*don*-ee-ee. With the
  swallows.

delavayi de-la-*vay*-ee. After Delavay.
dipterocarpum, dip-te-ro-*kar*-pum.
With a two-winged fruit.
flavum, fla-vum. Yellow. Yellow
Meadow Rue.
kiusianum, kee-oo-see-*a*-num. Of
Kyushu.
lucidum, loo-si-dum. Glossy.
orientale, o-ree-en-*ta*-lee. Eastern.

Thalictrum alpinum

Thelocactus, thee-lo-*kak*-tus.
Cactaceae. From Gk. *thele* (nipple)
and *Cactus.*
  bicolor, bi-ko-lor. Two-coloured
  (flowers).
  lophothele, lo-*fo*-thee-lee. With
  crested nipples.

Thermopsis, ther-*mop*-sis.
Leguminosae. From Gk. *thermos*
(lupin) and *-opsis* (resemblance).
Perennial herbs.
  caroliniana, ka-ro-lin-ee-*a*-na. Of
  Carolina. Carolina Lupin.
  montana, mon-*ta*-na. Of mountains.

Thlaspi, *thlas*-pee. *Cruciferae.* Gk. for
a cress. Perennial, alpine herb.
  arvense, ar-*ven*-see. In cultivated
  fields. Stinkweed.
  perfoliatum, per-fo-li-*a*-tum. With
  the leaf surrounding the stem.
  Pennycress.

rotundifolium, ro-tund-i-*fo*-lee-um.
With round leaves.

Thuja, *thoo*-ya. *Cupressaceae.* Gk.
name. Evergreen conifers. Thuja, Red
Cedar.
 koraiensis, ko-ree-*en*-sis. Of Korea.
 occidentalis, ok-si-den-*ta*-lis.
 Western. Western White Cedar.
 orientalis, o-ree-en-*ta*-lis. Eastern.
 plicata, pli-*ka*-ta. Plaited. Western
 Red Cedar.

Thujopsis, thoo-*yop*-sis.
Cupressaceae. From *Thuja* and Gk. -
opsis (resemblance). Evergreen
conifer.
 dolabrata, do-la-*bra*-ta. Hatchet-
 shaped (leaves).

Thunbergia, thun-*berg*-ee-a.
Acanthaceae. After Carl Peter
Thunberg. Tender perennial herbs,
shrubs and climbers.
 alata, a-*la*-ta. Winged. Black-eyed
 Susan.
 coccinea, kok-*kin*-ee-a. Scarlet.
 grandiflora, grand-i-*flo*-ra. Large-
 flowered. Blue Trumpet Vine.
 gregorii, gre-*go*-ree-ee. After Dr J.
 W. Gregory.
 myosorensis, mie-o-sor-*ren*-sis. Of
 Mysore.

Thymus, *tie*-mus. *Labiatae.* Gk. name.
Evergreen perennial herbs and shrubs.
Thyme.
 caespitosus, see-spi-*to*-sus. Tufted.
 carnosus, kar-*no*-sus. Fleshy.
 x *citriodorus,* sit-ree-o-*do*-rus.
 Lemon-scented. Lemon Thyme.
 herba-barona, her-ba-ba-*ron*-a.
 Corsican name. Caraway Thyme.
 nitidus, *nit*-id-us. Shining.
 praecox, *pree*-kox. Early.
 serpyllum, ser-*pil*-lum. L. for thyme.

Thymus serpyllum

Wild Thyme.
 vulgaris, vul-*ga*-ris. Common.
 Garden Thyme.

Tiarella, tee-a-*rel*-la. *Saxifragaceae.*
From Gk. *tiara* (small crown).
Perennial herbs. False Mitrewort.
 cordifolia, kor-di-*fo*-lee-a. With
 heart-shaped leaves. Foam Flower.
 trifoliata, tri-fo-lee-*a*-ta. With three
 leaves.
 wherryi, *we*-ree-ee. After Edgar
 Theodore Wherry.

Tibouchina, ti-boo-*chee*-na.
Melastomataceae. From the Guianan
name. Tender, evergreen perennial
herbs and shrubs.
 semidecandra, sem-i-dek-*an*-dra.
 Half ten-anthered.
 urvilleana, ur-vil-ee-*a*-na. After
 Jules Sebastian d'Urville. Purple
 Glory Tree.

Tigridia, tie-*gri*-dee-a. *Iridaceae.*
From L. *tigris* (tiger). Semi-hardy, bulbous herbs. Tiger Flower.
 pavonia, pa-*vo*-nee-a. Peacock-like.

Tilia, *ti*-lee-a. *Tiliaceae.* L. name.
Deciduous trees. Lime, Linden.

americana, a-me-ri-*ka*-na. Of America. American Lime.
cordata, kor-*da*-ta. Heart-shaped (leaves). Small-leaved Lime.
x *euchlora,* ew-*klo*-ra. Dark green. Caucasian Lime.
x *europaea,* ew-ro-*pee*-a. European. Common Lime.
mongolica, mon-*go*-lik-a. Of Mongolia.
oliveri, o-*li*-va-ree. After Oliver.
platyphyllos, pla-ti-*fil*-los. Broad-leaved. Broad-leaved Lime.
tomentosa, to-men-*to*-sa. Hairy (under the leaves). Silver Lime.

Tillandsia, ti-*land*-zee-a.
Bromeliaceae. After Elias Tillands.
Tender, evergreen epiphytic herbs.
argentea, ar-*jen*-tee-a. Silvery.
caput-medusae, *ka*-put-mee-*dew*-see. Like Medusa's head.
cyanea, sie-*an*-ee-a. Blue (flowers).
lindenii, lin-*den*-ee-ee. After J. J. Linden.
recurvata, re-*kur*-va-ta. Curved downwards.
stricta, *strik*-ta. Upright.
tenuifolia, ten-ew-i-*fo*-lee-a. Slender-leaved.
usneoides, us-nee-*oi*-deez. Hanging from trees. Spanish Moss.

Tolmiea, tol-*mee*-a. *Saxifragaceae.* After Dr William Tolmie. Perennial herb. Piggyback Plant.
menziesii, men-*zeez*-ee-ee. After Menzies.

Torenia, to-*ren*-ee-a.
Scrophulariaceae. After the Rev. Olof Toren. Annual and perennial herbs.
asiatica, a-see-*a*-ti-ka. Of Asia.
baillonii, bay-*lon*-ee-ee. After Henri Baillon.
fournieri, foor-nee-*e*-ree. After

Eugène Fournier. Bluewings.

Torreya, to-*ree*-a. *Cephalotaxaceae.* After John Torrey. Evergreen trees and shrubs.
californica, ka-li-*forn*-i-ka. Of California. California Nutmeg.
grandis, *grand*-is. Large. Chinese Nutmeg.
nucifera, new-*si*-fe-ra. Nut-bearing. Japanese Nutmeg.

Tovara, to-*va*-ra. *Polygonaceae.* After Simon Tovar. Perennial herb.
virginiana, vir-jin-ee-*a*-na. Of Virginia.

Townsendia, town-*zend*-ee-a.
Compositae. After David Townsend. Evergreen perennial and biennial herbs.
alpina, al-*pie*-na. Alpine.
exscapa, ex-*ska*-pa. Without a scape. Easter Daisy.
formosa, for-*mo*-sa. Beautiful.
grandiflora, gran-di-*flo*-ra. With large flowers.
parryi, pa-ree-ee. After Charles Christopher Parry.

Trachelium, tra-*kee*-lee-um.
Campanulaceae. From Gk. *trachelos* (neck). Perennial herbs.
caeruleum, see-*ru*-lee-um. Dark Blue. Throatwort.

Trachelospermum, tra-kee-lo-*sperm*-um. *Apocynaceae.* From Gk. *trachelos* (neck) and *sperma* (seed), after the narrow seeds. Evergreen climbers.
asiaticum, a-see-*a*-ti-kum. Asian.
jasminoides, jas-min-*oi*-deez. Jasmine-like.

Trachycarpus, tra-kee-*kar*-pus.
Palmae. From Gk. *trachys* (rough) and

karpos (fruit). Evergreen palm.
  fortunei, for-*tewn*-ee-ee. After
  Robert Fortune. Chusan Palm,
  Windmill Palm.

Trachymene, tra-kee-*mee*-nee.
Umbelliferae. From Gk. *trachys*
(rough) and *meninx* (membrane. Semi-
hardy, annual herb.
  caerulea, see-*ru*-lee-a. Dark Blue.
  Blue Lace Flower.

Tradescantia, tra-des-*kant*-ee-a.
Commelinaceae. After John
Tradescant. Tender and hardy perennial herbs. Spider Lily.
  albiflora, al-bi-*flo*-ra. White-flow-
  ered.
  blossfeldiana, blos-feld-ee-*a*-na.
  After Robert Blossfeld.
  fluminensis, floo-min-*en*-sis.
  Growing in a river. Speedy Jenny.
  navicularis, na-vik-ew-*la*-ris. Boat-
  shaped (leaves).
  sillamontana, si-la-mon-*ta*-na. Of
  Cerro de la Silla.

Tragopogon porrifolius

Tragopogon, tra-go-*po*-gon.
Compositae. From Gk. *tragos* (goat)
and *pogon* (beard). Biennial and
perennial herb. Goat's Beard.
  pratensis, pra-*ten*-sis. Of the meadows. Goat's Beard.
  porrifolius, po-ri-*fo*-lee-us. With
  leaves like *Allium porrum.* Salsify.

Trapa, *tra*-pa. *Trapaceae.* From L.
calcitrappa (four-pointed weapon).
Perennial, aquatic herbs.
  natans, na-tanz. Floating. Water
  Chestnut.

Trichocereus, tri-ko-*see*-ree-us.
Cactaceae. From Gk. *trichos* (hair)
and *Cereus,* after the hairy areoles.
Perennial cacti.
  bridgesii, bri-*jez*-ee-ee. After
  Thomas Bridges.
  candicans, kan-di-kanz. White
  (flowers).
  spachianus, spach-ee-*a*-nus. After
  Edouard Spach.

Tricyrtis, tri-*ser*-tis. *Liliaceae.* From
Gk. *tri* (three) and *kyrtos* (humped),
the swollen bases of the three outer
petals. Perennial herbs. Toad Lily.
  formosana, for-mo-*sa*-na. Of
  Formosa.
  hirta, hir-ta. Hairy. Toad Lily.
  macrantha, ma-*kranth*-a. Large-
  flowered.
  macropoda, ma-*kro*-po-da. With a
  large stalk.
  stolonifera, sto-lon-*iff*-er-a. Having
  stolons.

Trifolium, tri-*fo*-lee-um.
Leguminosae. From L. *tri-* (three) and
folium (leaf). Perennial herbs. Clover.
  alpinum, al-*pie*-num. Alpine.
  dubium, dub-ee-um. Dubious.
  Shamrock.

Trifolium repens

repens, ree-penz. Creeping. White Clover.
uniflorum, ew-ni-*flo*-rum. With one flower.

Trillium, *tri*-lee-um. *Liliaceae.* From L. *tri-* (three). The leaves and other parts are in threes. Perennial herbs. Wood Lily.
albidum, al-bi-dum. Whitish.
cernuum, ser-new-um. Nodding (flowers).
erectum, e-*rek*-tum. Erect (flowers). Birthroot.
grandiflorum, grand-i-*flo*-rum. Large-flowered. Wake Robin.
nivale, niv-*a*-lee. Snow-white.
ovatum, o-*va*-tum. Ovate (leaves).
rivale, ri-*va*-lee. Growing by streams.
sessile, se-si-lee. Sessile (flowers). Toadshade.
undulatum, un-dew-*la*-tum. Wavy-edged (petals). Painted Wood Lily.

Triteleia, tri-te-*lee*-a. *Liliaceae.* From Gk. *tri* (three) and *teleios* (perfect).

Cormous, perennial herbs.
hyacinthina, hi-a-sinth-*ee*-na. Hyacinth-coloured.
ixioides, ix-ee-*oi*-deez. *Ixia*-like.
laxa, lax-a. Loose (inflorescence). Triplet Lily.
peduncularis, pe-dunk-ew-*la*-ris. With a flower stalk.
uniflora, ew-ni-*flo*-ra. With one flower.

Tritonia, tri-*to*-nee-a. *Iridaceae.* From *Triton* (weather cock). Semi-hardy cormous, perennial herbs.
crocata, kro-*ka*-ta. Saffron yellow.
flavida, fla-vi-da. Yellow.
rosea, ro-see-a. Rose-coloured.

Trochodendron, tro-ko-*den*-dron. *Trochodendraceae.* From Gk. *trochos* (wheel) and *dendron* (tree), after the wheel-like flowers. Evergreen tree.
aralioides, a-ra-lee-*oi*-deez. *Aralia*-like.

Trollius, *tro*-lee-us. *Ranunculaceae.* From German *Trollblume* (Globe Flower). Perennial herbs. Globe Flower.
acaulis, a-*kaw*-lis. Stemless.
europaeus, ew-ro-*pee*-us. European. Globe Flower.
pumilus, pew-mi-lus. Dwarf.
yunnanensis, yoo-nan-*en*-sis. Of Yunnan, China.

Tropaeolum, tro-*pie*-o-lum. *Tropaeolaceae.* From Gk. *tropaion* (trophy). Hardy and semi-hardy, annual and perennial herbs. Nasturtium.
azureum, a-*zew*-ree-um. Blue (flowers).
canariense, ka-na-ree-*en*-see. Of the Canary Islands.
majus, ma-jus. Larger. Nasturtium.
peltophorum, pel-*to*-fo-rum. Shield

bearing (leaves).
peregrinum, pe-re-*green*-um.
Wandering. Canary Creeper.
polyphyllum, po-li-*fil*-lum. With
many leaves.
speciosum, spes-ee-*o*-sum. Showy.
Flame Nasturtium.
tricolorum, tri-kol-*or*-um. Three-
coloured.
tuberosum, tew-be-*ro*-sum.
Tuberous.

Tsuga, *tsoo*-ga. *Pinaceae.* From the
Japanese name. Evergreen trees.
Hemlock.
    canadensis, kan-a-*den*-sis. Of
Canada. Eastern Hemlock.
    diversifolia, die-ver-si-*fol*-ee-a.
Diversely leaved. Northern Japanese
Hemlock.
    heterophylla, he-te-ro-*fil*-la. With
variable leaves. Western Hemlock.
    mertensiana, mer-tenz-ee-*a*-na. After
Karl Heinrich Mertens. Mountain
Hemlock.
    sieboldii, see-*bold*-ee-ee. After
Siebold. Southern Japanese
Hemlock.

Tulipa, *tew*-li-pa. *Liliaceae.* From the
Turkish *tulband* (turban). Bulbous
perennial herbs. Tulip.
    acuminata, a-kew-mi-*na*-ta. Long-
pointed (petals). Horned Tulip.
    batalinii, ba-ta-*lin*-ee-ee. After A. F.
Batalin.
    biflora, bi-*flo*-ra. Two-flowered.
    clusiana, klooz-ee-*a*-na. After
Clusius. Lady Tulip.
    eichleri, *iek*-la-ree. After Eichler.
    fosteriana, fos-te-ree-*a*-na. After
Foster.
    greigii, *greeg*-ee-ee. After General
Greig.

Tulipa sylvestris

    humilis, *hu*-mi-lis. Low-growing.
    kaufmanniana, kowf-man-ee-*a*-na.
After General von Kaufmann.
    lanata, la-*na*-ta. Woolly.
    linifolia, lin-i-*fo*-lee-a. *Linum*-
leaved.
    orphanidea, or-fa-*nid*-ee-a. After Dr
Orphanides.
    polychroma, pol-ee-*kro*-ma. Many-
coloured.
    praestans, *pree*-stanz. Distinguished.
    pulchella, pul-*kel*-la. Pretty.
    saxatilis, sax-*a*-ti-lis. Growing
among rocks.
    sylvestris, sil-*ves*-tris. Of woods.
    wilsoniana, wil-son-ee-*a*-na. After
G. F. Wilson.

Typha, *tie*-fa. *Typhaceae.* The Gk.
name. Deciduous, perennial, aquatic
herbs. Bullrush.
    angustifolia, ang-gus-ti-*fo*-lee-a.
Narrow-leaved. Lesser Bullrush.
    latifolia, la-ti-*fo*-lee-a. Broad-leaved.
Cat's Tail.

U

Ulex, *ew*-lex. *Leguminosae.* L. name.
Almost leafless, spiny shrubs. Furze,
Gorse, Whin.
>*europaeus,* ew-ro-*pee*-us. European.
>Gorse.
>*minor,* *mi*-nor. Smaller. Dwarf
>Gorse.
>*parviflorus,* par-vi-*flo*-rus. Small-
>flowered.

Ulex europaeus

Ulmus, *ul*-mus. *Ulmaceae.* L. name.
Deciduous trees and shrubs. Elm.
>*americana,* a-me-ri-*ka*-na. Of
>America. White Elm.
>*angustifolia,* ang-gus-ti-*fo*-lee-a.
>Narrow-leaved. Goodyer's Elm.
>*canescens,* ka-*nes*-enz. Greyish-
>white hairs.
>*carpinifolia,* kar-pin-i-*fo*-lee-a.
>*Carpinus*-leaved. Smooth-leaved
>Elm.
>*cornubiensis,* kor-new-bee-*en*-sis. Of
>Cornwall. Cornish Elm.
>*glabra,* *glab*-ra. Smooth. Scotch Elm.

>*hollandica,* ho-*land*-i-ka. Of
>Holland. Dutch Elm.
>*parvifolia,* par-vi-*fo*-lee-a. Small-
>leaved. Chinese Elm.
>*procera,* pro-*see*-ra. Tall. English
>Elm.
>*pumila,* *pew*-mi-la. Dwarf.
>Siberian Elm.
>*sarniensis,* sar-nee-*en*-sis. Of
>Guernsey. Jersey Elm.

Umbellularia, um-bel-ew-*la*-ree-a.
Lauraceae. From L. *umbella* (umbel).
Evergreen tree.
>*californica,* ka-li-*forn*-i-ka. Of
>California. Myrtle, California
>Laurel.

Urceolina, ur-see-*o*-li-na.
Amaryllidaceae. From L. *urceolus*
(small pitcher), after the flower shape.
Bulbous herb.
>*peruviana,* pe-roo-vee-*a*-na. Of Peru.

Utricularia vulgaris

Urginea, ur-*gin*-ee-a. *Liliaceae.* From the Arabic *Beni Urgin.* Bulbous herb.
maritima, ma-*ri*-ti-ma. Growing near the sea. Sea Onion.

Ursinia, ur-*si*-nee-a. *Compositae.* After Johannes Ursinus. Annual and evergreen perennial herbs and sub-shrubs.
anethoides, a-nee-*thoi*-deez. *Anethum*-like.
anthemoides, an-them-*oi*-deez. *Anthemis*-like.
chrysanthemoides, kris-an-them-*oi*-deez. *Chrysanthemum*-like.

Utricularia, ew-trik-ew-*la*-ree-a. *Lentibulariaceae.* From L. *utriculus* (small bottle). Deciduous or evergreen carnivorous, aquatic herbs. Bladderwort.
exoleta, ex-o-*lee*-ta. Mature.
vulgaris, vul-*ga*-ris. Common.

Uvularia, ew-vew-*la*-ree-a. *Liliaceae.* From L. *uvula.* Perennial herbs. Bellwort, Wild Oats.
grandiflora, grand-i-*flo*-ra. Large-flowered.
perfoliata, per-fo-lee-*a*-ta. With the base of the leaves pierced by the stem.

V

Vaccinium, vax-*in*-ee-um. *Ericaceae.*
L. name. Deciduous or evergreen
shrubs.

angustifolium, an-gust-i-*fo*-lee-um.
Narrow-leaved.

arctostaphylos, ark-to-*sta*-fi-los.
Grapes eaten by bears. Caucasian
Whortleberry, Bearberry.

corymbosum, ko-rim-*bo*-sum. With
flowers in corymbs. Blueberry.

delavayi, de-la-*vay*-ee. After
Delavay.

deliciosum, dee-li-see-*o*-sum.
Delicious.

glaucoalbum, glow-ko-*al*-bum.
Glaucous-white (under the leaves).

myrtillus, *mur*-ti-lus. Small myrtle.
Bilberry, Whortleberry.

nummularia, num-ew-*la*-ree-a. Coin-
shaped (leaves).

oxycoccus, ox-ee-*kok*-us. Sharp-tasting berry. Small Cranberry.

vitis-idaea, vee-tis-ie-*dee*-a. Grape
of Mt Ida. Cowberry, Cranberry.

Vaccinium vitis-idaea

Valeriana, va-le-ree-*a*-na.
Valerianaceae. From L. *valere* (be
healthy), after its alleged healing properties. Perennial herbs.

montana, mon-*ta*-na. Of mountains.

officinalis, o-fi-si-*na*-lis. Sold in
shops. Common Valerian.

phu, foo. Evil-smelling.

saxatilis, sax-*a*-ti-lis. Growing among
rocks.

supina, su-*peen*-a. Prostrate.

Valeriana officinalis

Vallisneria, va-lis-*ne*-ree-a.
Hydrocharitaceae. After Antonio
Vallisnera. Evergreen, perennial,
aquatic herbs. Eel Grass.

americana, a-me-ri-*ka*-na. Of America.

gigantea, ji-*gan*-tee-a. Very large.

spiralis, spi-*ra*-lis. Spiralled. Eel
Grass.

Vallota, va-*lo*-ta. *Amaryllidaceae.*
After Pierre Vallot. Bulbous, perennial
herbs.
   speciosa, spes-ee-*o*-sa. George Lily,
   Scarborough Lily.

Vancouveria, van-koo-*ve*-ree-a.
Berberidaceae. After Captain George
Vancouver. Perennial and evergreen
herbs.
   chrysantha, kris-*anth*-a. With golden
   flowers.
   hexandra, hex-*an*-dra. With six stamens.

Vanda, *van*-da. *Orchidaceae.* From
the Hindi name. Greenhouse orchids.
   caerulea, see-ru-*lee*-a. Dark blue.
   caerulescens, see-ru-*les*-enz. Bluish.
   rothschildiana, roths-child-ee-*a*-na.
   After Lionel Walter, 2nd Baron
   Rothschild.
   teres, te-reez. Cylindrical.
   tricolor, tri-ko-lor. Three-coloured.

Veltheimia, vel-*tie*-mee-a. *Liliaceae.*
After August Ferdinand von Veltheim.
Tender, bulbous herbs.
   bracteata, brak-tee-*a*-ta. With bracts.
   capensis, ka-*pen*-sis. Of the Cape of
   Good Hope.
   viridiflora, vi-ri-di-*flo*-ra. Green-
   flowered. Forest Lily.

Venidium, vee-*ni*-dee-um.
Compositae. From L. *vena* (vein).
Semi-hardy annual and perennial
herbs.
   decurrens, dee-*ku*-renz. The leaf
   base merges with the stem.
   fastuosum, fas-tew-*o*-sum. Proud.

Veratrum, vee-*ra*-trum. *Liliaceae.*
The L. name. Perennial herbs. False
Hellebore.
   album, al-bum. White.

   californicum, ka-li-*forn*-i-kum. Of
   California.
   nigrum, nig-rum. Black.
   viride, vi-ri-dee. Green.

Verbascum, ver-*bas*-kum.
Scrophulariaceae. L. name. Biennial
and perennial herbs and sub-shrubs.
Mullein.
   blattaria, bla-*ta*-ree-a. *From* L.
   blatta (moth). Moth Mullein.
   chaixii, shay-zee-ee. After
   Dominique Chaix. Nettle-leaved
   Mullein.
   densiflorum, den-si-*flo*-rum. Densely
   flowered.
   dumulosum, dew-mew-*lo*-sum. Like
   a small shrub.
   nigrum, nig-rum. Black. Dark
   Mullein.
   olympicum, o-lim-pi-kum. Of Mt
   Olympus.
   orientale, o-ree-en-*ta*-lee. Eastern.
   phoeniceum, fee-*nee*-see-um.
   Purple-red. Purple Mullein.
   spinosum, spi-*no*-sum. Spiny.
   virgatum, vir-*ga*-tum. Wand-like.
   Twiggy Mullein.

Verbascum blattaria

Verbena officinalis

Verbena, ver-*be*-na. *Verbenaceae*. L. name. Biennial and perennial herbs. Vervain.
   alpina, al-*pie*-na. Alpine.
   bonariensis, bo-na-ree-*en*-sis. Of Buenos Aires. Purple Top.
   x *hybrida, hi*-bri-da. Hybrid. Florist's Verbena.
   officinalis, o-fi-si-*na*-lis. Sold in shops. Common Vervain.
   peruviana, pe-roo-vee-*a*-na. Of Peru.
   rigida, ri-ji-da. Rigid. Lilac Veined Verbena.
   tenera, te-ne-ra. Tender, delicate.

Vernonia, ver-*non*-ee-a. *Compositae*. After William Vernon. Annual and perennial herbs, trees and shrubs. Ironweed.
   altissima, al-*ti*-si-ma. Tallest.
   brasiliana, bra-zil-ee-*a*-na. Of Brazil.
   crinita, kri-*nee*-ta. Long-haired.
   flexuosa, flex-ew-*o*-sa. Tortuous.
   noveboracensis, no-vee-bor-a-*sen*-sis. Of New York.

Veronica, ve-*ro*-ni-ka. *Scrophulariaceae*. After St Veronica.

Perennial herbs and sub-shrubs. Speedwell, Bird's-Eye.
   alpina, al-*pie*-na. Alpine.
   austriaca, aw-stree-*a*-ka. Of Austria.
   cinerea, si-*ne*-ree-a. Grey.
   exaltata, ex-al-*ta*-ta. Very tall.
   fruticans, froo-ti-kanz. Shrubby. Rock Speedwell.
   gentianoides, jen-tee-a-*noi*-deez. *Gentiana*-like.
   incana, in-*ka*-na. Grey-hairy. Silver Speedwell.
   longifolia, long-i-*fo*-lee-a. Long-leaved.
   pectinata, pek-ti-*na*-ta. Comb-like (leaves).
   perfoliata, per-fo-lee-*a*-ta. With the leaf surrounding the stem.
   prostrata, pros-*tra*-ta. Prostrate.
   spicata, spee-*ka*-ta. With flowers in spikes.
   teucrium, tewk-ree-um. *Teucrium*-like.
   virginica, vir-*jin*-i-ka. Of Virginia.

Vestia, ves-*ti*-a, Solanaceae. After L.C. de Vest. Evergreen shrub.
   foetida, fee-ti-da. Foetid.

Viburnum, vee-*bur*-num. *Caprifoliaceae*. L. name. Deciduous and evergreen trees and shrubs. Arrow Wood.
   x *bodnantense,* bod-nant-*en*-see. Of Bodnant.
   x *burkwoodii,* burk-*wud*-ee-ee. After Albert Burkwood.
   carlesii, karlz-ee-ee. After W. R. Carles.
   cinnamomifolium, sin-a-mo-mi-*fo*-lee-um. *Cinnamom*-leaved.
   davidii, da-*vid*-ee-ee. After David.
   farreri, fa-ra-ree. After Reginald Farrer.
   fragrans, fra-granz. Fragrant.
   grandiflorum, grand-i-*flo*-rum.

Large-flowered.

henryi, hen-ree-ee. After Henry.

x *juddii, jud*-ee-ee. After William H. Judd.

lantana, lan-*ta*-na. L. name for *Viburnum.* Twistwood.

opulus, op-ew-lus. L. name for maple. Guelder Rose, Crampbark.

plicatum, pli-*ka*-tum. Pleated (leaves).

sargentii, sar-*jent*-ee-ee. After Sargent.

tinus, teen-us. The L. name. Laurustinus.

trilobum, tri-*lo*-bum. With three lobes. Cranberry.

Vicia, *vi*-see-a. *Leguminosae.* From L. for vetch. Annual and perennial herb. Vetch, Tare.

faba, fa-ba. The L. name. Broad Bean.

Victoria, vik-*tor*-ree-a, *Nymphaeaceae.* After Queen Victoria. Tender, aquatic perennial herbs. Giant Water Lily.

amazonica, a-ma-*zon*-i-ka. Of the Amazon. Amazon Water Lily, Royal Water Lily.

cruziana, krooz-ee-*a*-na. Of Santa Cruz. Santa Cruz Water Lily.

Vinca, *vin*-ca, *Apocynaceae.* From L. *vincio* (bind). Evergreen sub-shrubs and perennials. Periwinkle.

difformis, di-*for*-mis. Of dissimilar shapes.

major, ma-jor. Larger. Greater Periwinkle.

minor, mi-nor. Smaller. Lesser Periwinkle.

rosea, ro-see-a. Rose-coloured.

Viola, *vie*-o-la, *Violaceae.* L. for scented flowers. Annual and perennial herbs

and deciduous sub-shrubs. Violet.

aetolica, ee-*tol*-i-ka. Of Aitolia, Greece.

alba, al-ba. White.

biflora, bi-*flo*-ra. Two-flowered.

blanda, blan-da. Pleasant. Sweet White Violet.

canina, kan-*ee*-na. Pertaining to dogs. Dog Violet.

cornuta, kor-*new*-ta. Horned (long spur). Viola, Bedding Pansy.

cucullata, kuk-ew-*la*-ta. Hood-like.

gracilis, gra-si-lis. Graceful.

hederacea, he-de-*ra*-see-a. *Hedera*-like (leaves). Trailing Violet.

labradorica, lab-ra-*do*-ri-ka. Of Labrador. Labrador Violet.

lutea, loo-tee-a. Yellow. Mountain Pansy.

odorata, o-do-*ra*-ta. Scented. Sweet Violet, English Violet.

Viola odorata

palmata, parl-*ma*-ta. Lobed like a hand. Wild Okra.

pedata, pe-*da*-ta. Like a bird's foot (leaves). Bird's-foot Violet, Pansy Violet.

sempervirens, sem-per-*vi*-rens. Evergreen. Evergreen Violet.

septentrionalis, sep-ten-tree-o-*na*-lis. Northern.

tricolor, tri-ko-lor. Three-coloured.
Wild Pansy, Heartsease.
x *wittrockiana,* wit-rok-ee-*a*-na.
After Professor Veit Wittrock. Pansy.

Viscum, *vis*-kum. *Viscaceae*. L. name.
Evergreen parasitic shrub.
  album, al-bum. White (fruit).
  Mistletoe.

Viscum album

Vitaliana, vi-ta-lee-*a*-na. *Primulaceae*.
After Vitaliano Donati. Evergreen
perennial herb.
  primuliflora, prim-ew-li-*flo*-ra.
  Primula-flowered.

Vitex, *vi*-tex. *Verbenaceae*. L. name.
Evergreen or deciduous shrubs.
  agnus-castus, ag-nus-*kas*-tus. L. for
  Chaste-Tree.
  capitata, ka-pi-*ta*-ta. In a dense head.
  negundo, ne-*gun*-do. Native name.

Vitis, *vee*-tis. *Vitaceae*. L. for grape
vine. Deciduous climbers. Vine.
  aconitifolia, a-kon-ee-ti-*fo*-lee-a.
  Aconitum-leaved.
  amurensis, am-ew-*ren*-sis. Of the
  Amur River region. Amur Grape.
  quinquefolia, kwin-kwee-*fo*-lee-a.
  With five leaves.
  striata, stri-*a*-ta. Striped.
  vinifera, veen-*i*-fe-ra. Wine-bearing.
  Common Grape Vine.

Vriesia, *vree*-zee-a. *Bromeliaceae*.
After Willem de Vriese. Tender, evergreen and perennial herbs.
  fenestralis, fe-ne-*stra*-lis. Window-
  like.
  fosteriana, fos-ta-ree-*a*-na. After
  Milford Foster.
  hieroglyphica, hi-e-ro-*gli*-fi-ka.
  Marked with hieroglyphs (leaves).
  psittacina, si-ta-*seen*-a. Parrot-like.
  regina, ree-*jeen*-a. Queen.
  splendens, splen-denz. Splendid.
  Flaming Sword.

W

Wahlenbergia, war-lan-*berg*-ee-a.
Campanulaceae. After Georg
Wahlenberg. Biennial or perennial
herbs. Rock Bell.
  albomarginata, al-bo-mar-ji-*na*-ta.
  White-margined.
  congesta, con-*jes*-ta. Congested.
  gracilis, gra-si-lis. Graceful.
  hederacea, he-de-*ra*-see-a. *Hedera*-
  like (leaves).
  matthewsii, math-*ewz*-ee-ee. After H.
  J.Matthews.
  saxicola, sax-*i*-ko-la. Growing on
  rocks.

Waldsteinia, wald-*stien*-ee-a.
Rosaceae. After Count Franz
Waldstein-Wartenburg. Creeping
perennial herbs.
  fragarioides, fra-ga-ree-*oi*-deez.
  Fragaria-like. Barren Strawberry.
  ternata, ter-*na*-ta. In groups of three.
  trifolia, tri-*fo*-lee-a. With three
  leaves.

Washingtonia, wosh-ing-*ton*-ee-a.
Palmae. After President George
Washington. Tender, evergreen palms.
  filifera, fil-i-fe-ra. Thread-bearing.
  Cotton Palm.
  robusta, ro-*bus*-ta. Robust. Southern
  Washingtonia.

Watsonia, wot-*son*-ee-a. *Iridaceae*.
After Sir William Watson. Semi-hardy
perennial herbs.
  amabilis, a-*ma*-bi-lis. Beautiful.
  beatricis, bee-*a*-tri-sis. After Beatrix
  Hops.
  fourcadei, foor-*kad*-ee-ee. After H.
  G. Fourcade.

  meriana, me-ree-*a*-na. After Sybilla
  Merian.
  pyramidata, pi-ra-mi-*da*-ta. Pyramidal.
  tabularis, tab-ew-*la*-ris. Of Table
  Mountain.

Weigela, *wie*-ge-la. *Caprifoliaceae*.
After Christian von Weigel. Deciduous
shrubs.
  decora, de-*ko*-ra. Beautiful.
  florida, flo-ri-da. Flowering.
  praecox, pree-kox. Early flowering.

Wisteria, wis-*te*-ree-a. *Leguminosae*.
After Caspar Wistar. Deciduous, twining climbers.
  chinensis, chin-*en*-sis. Of China.
  floribunda, flo-ri-*bun*-da. Profusely
  flowering. Japanese Wisteria.
  formosa, for-*mo*-sa. Beautiful.
  sinensis, si-*nen*-sis. Of China.
  Chinese Wisteria.
  venusta, ven-*us*-ta. Handsome. Silky
  Wisteria.

Wulfenia, wul-*fen*-ee-a.
Scrophulariaceae. After Franz von
Wulfen. Evergreen perennial herbs.
  baldaccii, bal-*dak*-ee-ee. After
  Antonio Baldacci.
  carinthiaca, ka-rinth-ee-*a*-sa. Of
  Carinthia, Austria.
  orientalis, o-ree-en-*ta*-lis. Eastern.

Woodwardia, wood-*ward*-ee-a.
Blechnaceae. After Mr. T. J.
Woodward. Semi-hardy, evergreen or
deciduous ferns. Chain Fern.
  radicans, ra-di-kanz. With rooting
  stems. Chain Fern
  virginica, vir-*jin*-i-ka. Of Virginia.

X

Xantheranthemum, zanth-e-*ranth*-e-mum. *Acanthaceae.* From Gk. *xanthos* (yellow) and *Eranthemum.* Tender perennial herb.
 igneum, ig-nee-um. Fiery red.

Xanthoceras, zanth-*o*-se-ras. *Sapindaceae.* From Gk. *xanthos* (yellow) and *keras* (horn). Deciduous trees or shrubs.
 sorbifolium, sor-bi-*fo*-lee-um. *Sorbus*-leaved.

Xanthorhiza, zanth-o-*reez*-a. *Ranunculaceae.* From Gk. *xanthos* (yellow) and *rhiza* (root). Deciduous shrub.
 simplicissima, sim-pli-*si*-si-ma. Most simple. Yellow Root.

Xanthosoma, zan-tho-*so*-ma. *Araceae.* From Gk. *xanthos* (yellow) and *soma* (body). Tender, perennial tuberous herbs.
 x *sagittifolium,* sa-ji-ti-*fo*-lee-um. With arrow-shaped leaves.
 x *violaceum,* vie-o-*la*-see-um. Violet.

Xeranthemum, ze-*ranth*-e-mum. *Compositae.* From Gk. *xeros* (dry) and *anthos* (flower). Semi-hardy, annual herb.
 x *annuum, an*-ew-um. Annual. Immortelle.

Xerophyllum, ze-ro-*fil*-um. *Liliaceae.* From Gk. *xeros* (dry) and *phyllon* (leaf). Perennial herbs.
 x *tenax, ten*-ax. Tough. Elk Grass.

Y

Yucca, *yew*-ka. *Agavaceae.* The Caribbean name for Cassava. Tender to hardy, evergreen trees and shrubs.
 aloifolia, a-lo-i-*fo*-lee-a. *Aloe*-leaved. Dogger Plant.
 filamentosa, fil-a-men-*to*-sa. With filaments. Needle Palm.
 flaccida, fla-si-da. Flaccid (leaves).
 gloriosa, glo-ree-*o*-sa. Glorious.

Palm Lily.
 parviflora, par-vi-*flo*-ra. Small-flowered.
 smalliana, small-ee-*a*-na. After John Kunkel Small. Bear Grass.
 whipplei, wi-pal-ee. After Lieut. Amiel Weeks Whipple. Our Lord's Candle.

Z

Zantedischia, zan-te-*di*-see-a.
Araceae. After Francesco Zantedischi.
Tender to hardy, evergreen, tuberous
perennial herbs. Arum Lily, Calla Lily.
  aethiopica, ee-thee-*o*-pi-ka. Of
  Africa. Arum Lily.
  albomaculata, al-bo-mak-ew-*la*-ta.
  White-spotted (leaves).
  elliottiana, e-lee-o-tee-*a*-na. After
  Elliott.
  rehmannii, ree-*man*-ee-ee. After
  Rehmann.

Zanthoxylum, zanth-*ox*-i-lum.
Rutaceae. From Gk. *xanthos* (yellow)
and *xylon* (wood). Deciduous or
evergreen trees and shrubs. Prickly
Ash.
  ailanthoides, ie-lanth-*oi*-deez.
  Ailanthus-like.
  americanum, a-me-ri-*ka*-num. Of
  America. Toothache Tree.
  piperitum, pi-pe-*ree*-tum. Pepper-
  like (seed taste). Japan Pepper.
  planispinum, pla-ni-*speen*-um. With
  flat spines.
  schinifolium, skeen-i-*fo*-lee-um.
  With *Schinus*-like leaves.
  simulans, sim-*ew*-lans. Resembling.

Zauschneria, zow-*shne*-ree-a.
Onagraceae. After Johann Baptist
Zauschner. Perennial sub-shrubs.
  californica, ka-li-*forn*-i-ka. Of
  California.
  cana, *ka*-na. Grey.

Zea, *zee*-a. *Gramineae*. From Gk.
Annual grass.
  mays, mayz. From the Mexican
  name. Maize, Corn, Sweet Corn.

  gracillima, gra-sil-*ee*-ma. Most
  graceful.

Zelkova, zel-*ko*-va. *Ulmaceae*. From
the Caucasian name. Deciduous trees.
  carpinifolia, kar-pie-ni-*fo*-lee-a.
  Carpinus-leaved.
  serrata, se-*ra*-ta. Saw-toothed
  (leaves).

Zenobia, zen-*o*-bee-a. *Ericaceae*.
After Zenobia, a Queen of Palmyra.
Deciduous or semi-evergreen shrub.
  pulverulenta, pul-ve-ru-*len*-ta.
  Powdered.

Zephyranthes, ze-fi-*ranth*-eez.
Amaryllidaceae. From Gk. *zephyros*
(west wind) and *anthos* (flower).
Tender to semi-hardy, bulbous herbs.
Zephyr Flower, Rain Lily.
  atamasco, a-ta-*mas*-ko. A native
  name. Atamasco Lily.
  candida, kan-di-da. White.
  carinata, ka-ri-*na*-ta. Keeled.
  citrina, si-*tree*-na. Lemon-yellow.
  grandiflora, grand-i-*flo*-ra. Large-
  flowered.
  robusta, ro-*bus*-ta. Robust.
  rosea, ro-see-a. Rose-coloured.

Zigadenus, zi-ga-*dee*-nus. *Liliaceae*.
From Gk. *zygos* (yoke) and *aden*
(gland). Bulbous perennial herb.
Zigadene.
  elegans, e-le-ganz. Elegant. Alkali
  Grass.
  fremonti, free-*mont*-ee-ee. After
  Colonel John Freemont. Star Lily.

Zingiber, *zin*-ji-ber. *Zingiberaceae*.

From L. *zingiber* (ginger). Tender, perennial herbs.
officinale, o-fi-si-*na*-lee. Sold in shops. Ginger.
purpureum, pur-*pur*-ree-um. Purple. Bengal Ginger.

Zinnia, *zin*-ee-a. *Compositae.* After Johann Gottfried Zinn. Semi-hardy, annual herbs.
angustifolia, an-gus-ti-*fo*-lee-a. Narrow-leaved.
elegans, *e*-le-ganz. Elegant.
grandiflora, gran-di-*flo*-ra. With large flowers.
haageana, harg-ee-*a*-na. After F. A. Haage.

Zizania, zi-*zay*-ni-a. Gramineae. From Gk. *zizanion* (weed). Perennial aquatic grass.
aquatica, a-*kwa*-ti-ka. Growing in water. Canadian Wild Rice.
latifolia, la-ti-*fo*-lee-a. Broad-leaved. Water Rice.

Zygopetalum, zi-go-*pe*-ta-lum. *Orchidaceae.* From Gk. *zygos* (yoke) and *petalon* (petal). Greenhouse orchids.
crinitum, kree-*nee*-tum. Long-haired.
intermedium, in-ter-*me*-dee-um. Intermediate.
mackayi, ma-*kay*-ee. After Mr Mackay.

ENGLISH-LATIN PLANT NAMES

A

Aaron's Beard	*Hypericum calycinum*
Aaron's Rod	*Verbascum thapsus*
Abele	*Populus alba*
Abyssinian Feathertop	*Pennisetum villosum*
Acacia	*Acacia*
False	*Robinia pseudacacia*
Rose	*R. hispida*
Aconite	*Aconitum*
Winter	*Eranthis*
Adam's Needle	*Yucca filamentosa*
Adder's tongue	*Ophioglossum*
African Boxwood	*Myrsine africana*
African Corn-lily	*Ixia*
African Daisy	*Arctotis*
African Hemp	*Sparmannia africana*
African Lily	*Agapanthus*
African Violet	*Saintpaulia*
Agrimony	*Agrimonia*
Air Plant	*Kalanchoe pinnata*
Akee tree	*Blighia sapida*
Alder	*Alnus*
Common	*A. glutinosa*
Grey	*A. incana*
Italian	*A. cordata*
Japanese	*A. japonica*
Alder Buckthorn	*Rhamnus frangula*
Alecost	*Balsamita major*
Alexandrian Laurel	*Danae racemosa*
Alfalfa	*Medicago sativa*
Algerian Statice	*Limonium bonduellii*
Alice, Sweet	*Pimpinella anisum*
Allegheny Spurge	*Pachysandra procumbens*
Allspice, Californian	*Calycanthus occidentalis*
Carolina	*C. floridus*
Almond	*Prunus dulcis*
Dwarf Russian	*P. tenella*
Aluminium Plant	*Pilea cadierei*
Alum Root	*Heuchera*
Alyssum, sweet	*Lobularia maritima*
Anemone	*Anemone*
Poppy	*A. coronaria*
Snowdrop	*A. sylvestris*
Wood	*A. nemorosa*
Angel's Fishing Rod	*Dierama pulcherrimum*
Angel's Tears	*Billbergia nutans, Narcissus triandrus*
Golden	*Narcissus triandrus pallidulus*
Angel's Trumpet	*Datura arborea*

Angel's Wings	*Caladium*
Angelica, wild	*Angelica sylvestris*
Animated Oat	*Avena sterilis*
Anise	*Pimpinella anisum*
Annatto	*Bixa orellana*
Apple	*Malus*
Apple of Peru	*Nicandra physalodes*
Apricot	*Prunus armeniaca*
Japanese	*P. Mume*
Arrowhead	*Sagittaria*
Arrowhead Vine	*Syngonium angustatum*
Arrowroot	*Maranta arundinacea*
Artichoke, Chinese	*Stachys affinis*
Globe	*Cynara cardunculus*
Jerusalem	*Helianthus tuberosus*
Artillery Plant	*Pilea microphylla*
Arum-lily	*Zantedeschia*
Ash	*Fraxinus*
Arizona	*F. velutina*
Common	*F. excelsior*
Manna	*F. ornus*
Narrow-leaved	*F. angustifolia*
White	*F. americana*
Asoka tree	*Saraca indica*
Asparagus	*Asparagus officinalis*
Asparagus Pea	*Lotus tetragonolobus*
Aspen	*Populus tremula*
Asphodel	*Asphodelus*
White	*A. albus*
Yellow	*Asphodeline lutea*
Aster, China	*Callistephus Chinensis*
Stokes'	*Stokesia*
Astilbe	*Astilbe*
Aubergine	*Solanum melongena*
Auricula	*Primula auricula*
Australian Bluebell Creeper	*Sollya heterophylla*
Australian Honeysuckle	*Banksia*
Austrian Copper Briar	*Rosa foetida bicolor*
Avens	*Geum*
White	*G. canadense*
Wood	*G. urbanum*
Yellow	*G. aleppicum*
Avocado Pear	*Persea americana*

B

Baboon Root	*Babiana*
Baby Blue Eyes	*Nemophila*
Baby Rubber Plant	*Peperomia obtusifolia*

Baby's Breath	*Gypsophila paniculata*
Baby's Tears	*Hypoestes phyllostachya, Soleirolia soleirolii*
Bachelor's Buttons	*Tanacetum parthenium*
White	*Ranunculus aconitifolius*
Yellow	*R. acris flore pleno*
Badger's Bane	*Aconitum vulparia*
Bael Fruit	*Aegle marmelos*
Baldmoney	*Meum athamanticum*
Balloon Flower	*Platycodon grandiflorus*
Balloon Vine	*Cardiospermum halicacabum*
Balm, Bee	*Melissa officinalis*
Balm, Lemon	*Melissa officinalis*
Balm of Gilead	*Populus candicans*
Balsam	*Impatiens balsamina*
Himalayan	*I. glandulifera*
Balsam-apple	*Momordica balsamina*
Balsam-pear	*Momordica charantia*
Bamboo	*Arundinaria, Phyllostachys, Sasa*
Black	*Phyllostachys nigra*
Banana	*Musa*
Baneberry	*Actaea*
Red	*A. rubra*
Banyan Tree	*Ficus benghalensis*
Barbados Gooseberry	*Pereskia aculeata*
Barbados Pride	*Caesalpinia pulcherrima*
Barberry	*Berberis*
Common	*B. vulgaris*
Barberton Daisy	*Gerbera jamesonii*
Barley	*Hordeum*
Barrenwort	*Epimedium*
Basil	*Ocimum basilicum*
Bush	*O. minimum*
Basket Grass	*Oplismenus hirtellus*
Basket Plant	*Aeschynanthus*
Bat Plant	*Tacca integrifolia*
Bats-in-the-Belfry	*Campanula trachelium*
Bay	*Laurus*
Laurel	*L. nobilis*
Bayberry	*Myrica pensylvanica*
California	*M. californica*
Beach Grass	*Ammophila*
Beach Heather	*Hudsonia*
Beach Pea	*Lathyrus*
Bead Plant	*Nertera depressa*
Bead Tree	*Melia azederach*
Bearberry	*Arctostaphylos uva-ursi*

Beard Grass	*Polypogon monspeliensis*
Bear's Breeches	*Acanthus*
Beauty Berry	*Callicarpa*
Beauty Bush	*Kolkwitzia amabilis*
Bedstraw	*Galium*
Bee Balm	*Monarda didyma*
Beech	*Fagus*
Common	*F. sylvatica*
Copper	*F. sylvatica Purpurea*
Beefsteak Plant	*Iresine herbstii*
Beefwood	*Casuarina*
Beetroot	*Beta vulgaris*
Belladonna Lily	*Amaryllis belladonna*
Bellflower	*Campanula*
Adriatic	*C. garganica*
Chimney	*C. pyramidalis*
Clustered	*C. glomerata*
Giant	*C. latifolia, Ostrowskia magnifica*
Italian	*C. isophylla*
Milky	*C. lactiflora*
Spurred	*C. alliariifolia*
Bells of Ireland	*Moluccella laevis*
Bellwort	*Uvularia*
Benjamin Bush	*Lindera benzoin*
Bergamot	*Monarda didyma*
Big Tree	*Sequoiadendron giganteum*
Bilberry	*Vaccinium myrtillus*
(U.S.A.)	*V. uliginosum*
Bindweed	*Convolvulus Calystegia*
Birch	*Betula*
Dwarf	*B. nana*
Himalayan	*B. utilis*
Japanese Cherry	*B. grossa*
Paper	*B. papyrifera*
River	*B. nigra*
Silver	*B. pendula*
Sweet	*B. lenta*
Yellow	*B. lutea*
Bird of Paradise Flower	*Strelitzia reginae*
Bird's Eyes	*Gilia tricolor*
Bird's Foot Trefoil	*Lotus corniculatus*
Birthroot	*Trillium*
Birthwort	*Aristolochia*
Bishop's Cap	*Mitella*
Bishop's Weed	*Aegopodium podagraria*
Bishop's Wort	*Stachys macrantha*
Bistort	*Polygonum bistorta*

Blackberry	*Rubus*
Black-eyed Susan	*Rudbeckia hirta, Thunbergia alata*
Black Gum	*Nyssa sylvatica*
Black Snakeroot	*Cimicifuga racemosa*
Blackthorn	*Prunus spinosa*
Blackwood	*Acacia melanoxylon*
Bladder-nut	*Staphylea*
Bladder Senna	*Colutea arborescens*
Bladderwort	*Utricularia*
Blanket Flower	*Gaillardia*
Blazing Star	*Mentzelia lindleyi*
Bleeding Heart	*Dicentra spectabilis*
Bleeding-heart Vine	*Clerodendrum thomsoniae*
Blood Flower	*Haemanthus katharinae*
Blood Lily	*Haemanthus*
Bloodleaf	*Iresine herbstii*
Bloodroot	*Sanguinaria*
Bloodwood Tree	*Haematoxylum campechianum*
Blue-eyed Mary	*Collinsia verna, Omphalodes verna*
Blue Flowered Torch	*Tillandsia lindenii*
Blue Lace Flower	*Trachymene caerulea*
Blue Lips	*Collinsia grandiflora*
Blue Thimble Flower	*Gilia capitata*
Blue Trumpet Vine	*Thunbergia grandiflora*
Bluebell	*Hyacinthoides non-scripta*
(Scotland)	*Campanula rotundifolia*
Spanish	*Hyacinthoides hispanica*
Blueberry	*Vaccinium*
Box	*V. ovatum*
Highbush	*V. corymbosum*
Bluebush	*Kochia*
Bluets	*Hedyotis caerulea*
Blushing Bromeliad	*Neoregelia carolinae*
Boat Lily	*Rhoeo spathacea*
Bog Arum	*Calla palustris*
Bog Bean	*Menyanthes trifoliata*
Bog Myrtle	*Myrica gale*
Bog Rosemary	*Andromeda*
Boneset	*Eupatorium perfoliatum*
Borage	*Borago officinalis*
Borecole	*Brassica oleracea* Acephala
Boston Ivy	*Parthenocissus tricuspidata*
Bottle-brush	*Callistemon*
Crimson	*C. citrinus*
Bottle Gourd	*Lagenaria siceraria*
Bouncing Bet	*Saponaria officinalis*
Bower Plant	*Pandorea jasminoides*

Box	*Buxus*
Common	*B. sempervirens*
Box Elder	*Acer negundo*
Box Thorn	*Lycium barbarum*
Brake, Australian	*Pteris tremula*
Cretan	*P. cretica*
Sword	*P. ensiformis*
Brandy Bottle	*Nuphar lutea*
Brasiletto	*Caesalpina vesicaria*
Brass Buttons	*Cotula coronopifolia*
Brazilian Edelweiss	*Sinningia leucotricha*
Brazilian Plume	*Justicea carnea*
Brazil Nut	*Bertholletia*
Breadfruit	*Artocarpus altilis*
Bridal Wreath	*Francoa sonchifolia, Spiraea arguta*
Bridewort	*Spiraea salicifolia*
Broad Bean	*Vicia faba*
Broad-leaved Kindling Bark	*Eucalyptus dalrympleana*
Broccoli	*Brassica oleracea botrytis*
Sprouting	*B. oleracea italica*
Brooklime	*Veronica beccabunga*
Broom	*Cytisus, Genista*
Common	*Cytsus scoparius*
Dalmatian	*Genista sylvestris*
Genoa	*G. januensis*
Montpelier	*Cytisus monspessulanus*
Mt Etna	*Genista aetnensis*
Pineapple-scented	*Cytisus battandieri*
White Spanish	*C. multiflorus*
Brussels Sprout	*Brassica oleracea gemmifera*
Bryony	*Bryonia*
Black	*Tamus communis*
White	*Bryonia dioica*
Buck Bean	*Menyanthes trifoliata*
Buckeye	*Aesculus*
California	*A. Californica*
Red	*A. pavia*
Yellow	*A. Flava*
Buckthorn, Common	*Rhamnus cathartica*
Bugbane	*Cimicifuga*
Bugle	*Ajuga reptans*
Bugleweed	*Lycopus virginicus*
Bugloss	*Anchusa*
Viper's	*Echium*
Bullace	*Prunus spinosa insititia*
Bull Bay	*Magnolia grandiflora*
Bulrush	*Typha latifolia*

Bunny Rabbits	*Linaria maroccana*
Burdock	*Arctium*
Burhead	*Echinodorus*
Burning Bush	*Dictamnus albus, Kochia scoparia*
Burnet	*Sanguisorba*
Burstwort	*Herniaria*
Bush Clover	*Lespedeza*
Bush Groundsel	*Baccharis halimifolia*
Busy Lizzie	*Impatiens walleriana*
Butcher's Broom	*Ruscus aculeatus*
Butterbur	*Petasites*
Buttercup	*Ranunculus*
Persian	*R. Asiaticus*
Butterfly Bush	*Buddleia davidii*
Butterfly Flower	*Schizanthus*
Butterfly Lily	*Hedychium coronarium*
Butterfly Tree	*Bauchinia purpurea*
Butter-nut	*Juglans cinerea*
Butterwort	*Pinguicula*
Buttonwood	*Platanus occidentalis*
Button Snake Root	*Liatris pycnostachya*
Buttons-on-a-string	*Crassula rupestris*

C

Cabbage	*Brassica oleracea capitata*
Chinese	*B. rapa pekinensis*
Portuguese	*B. oleracea tronchuda*
Wild	*B. oleracea*
Cabbage Gum	*Eucalyptus pauciflora*
Cabbage Tree	*Cordyline australis*
Calaba Tree	*Calophyllum*
Calamondin	*Citrofortunella mitis*
Calico Bush	*Kalmia latifolia*
Calico Flower	*Aristolochia elegans*
Calico Hearts	*Adromischus maculatus*
California Bluebell	*Phacelia campanulata*
California Geranium	*Senecio petasites*
California Laurel	*Myrica californica*
California Nutmeg	*Torreya californica*
Campernelle Jonquil	*Narcissus x odorus*
Campion	*Silene*
Alpine	*Lychnis alpina*
Bladder	*Silene vulgaris*
Moss	*S. acaulis*
Rose	*Lychnis coronaria*
Sea	*Silene vulgaris maritima*
Canada Lily	*Lilium canadense*
Canada Ted	*Gaultheria*

Canary Creeper	*Tropaeolum speciosum*
Canary Grass	*Phalaris canariensis*
Candelabra Plant	*Aloe arborescens*
Candle Plant	*Plectranthus oertendahlii, Senecio articulatus*
Candlewick	*Verbascum thapsus*
Candytuft	*Iberis*
Cantaloupe	*Cucumis melo*
Canterbury Bells	*Campanula medium*
Cup and Saucer	*C. medium calycanthema*
Cape Asparagus	*Aponogeton distachyos*
Cape Cowslip	*Lachenalia*
Cape Gooseberry	*Physalis peruviana*
Cape Honeysuckle	*Tecomaria capensis*
Cape Ivy	*Senecio macroglossus*
Cape Jasmine	*Gardenia jasminoides*
Cape Leadwort	*Plumbago auriculata*
Cape Pondweed	*Aponogeton distachyos*
Caper	*Capparis spinosa*
Caper Spurge	*Euphorbia lathyris*
Carambola	*Averrhoa carambola*
Caraway	*Carum carvi*
Cardamon	*Elettaria cardamomum*
Cardinal Flower	*Lobelia cardinalis, Sinningia cardinalis*
Cardoon	*Cynara cardunculus*
Carnation	*Dianthus caryophyllus*
Carob	*Ceratonia siliqua*
Carolina Lupin	*Thermopsis caroliniana*
Carrot	*Daucus carota sativus*
Wild	*D. carota*
Cartwheel Flower	*Heracleum mantegazzianum*
Cashew Nut	*Anacardium occidentale*
Cassava	*Manihot*
Bitter	*M. esculenta*
Sweet	*M. dulcis*
Cast Iron Plant	*Aspidistra elatior*
Castor-oil Plant	*Ricinus communis*
False	*Fatsia japonica*
Cat Brier	*Smilax*
Cat Thyme	*Teucrium marum*
Catchfly, German	*Lychnis viscaria*
Nodding	*Silene pendula*
Nottingham	*S. nutans*
Cathedral Bells	*Cobaea scandens*
Cathedral Windows	*Calathea makoyana*
Catjang	*Vigna sinensis cylindrica*

Catmint	*Nepeta cataria*
Catnip	*Nepeta cataria*
Cat's Claw	*Doxantha ungus-cati*
Cat's Foot	*Antennaria dioica*
Cat's Tail	*Phleum, Typha*
Cat's Whiskers	*Tacca chantrieri*
Cauliflower	*Brassica oleracea botrytis*
Cedar	*Cedrus*
Atlas	*C. Atlantica*
Cedar of Goa	*Cupressus lusitanica*
Cedar of Lebanon	*Cedrus libani*
Celandine, Greater	*Chelidonium majus*
Lesser	*Ranunculus ficaria*
Celeriac	*Apium graveolens rapaceum*
Celery	*A. graveolens dulce*
Wild	*A. graveolens*
Celery Pine	*Phyllocladus alpinus*
Centaury	*Centaurium*
American	*Sabatia*
Century Plant	*Agave americana*
Cereus, night-blooming	*Selenicereus grandiflorus*
Chain Plant	*Tradescantia navicularis*
Chamomile	*Chamaemelum nobile*
Dyer's	*Anthemis tinctoria*
Stinking	*A. cotula*
Yellow	*A. tinctoria*
Chandelier Plant	*Kalanchoe tubiflora*
Chard, Swiss	*Beta vulgaris cicla*
Chaste Tree	*Vitex agnus-castus*
Chatham Island Forget-me-not	*Myostidium hortensia*
Chayote	*Sechium edule*
Chenille Plant	*Acalypha hispida*
Cherimoya	*Anona cherimola*
Cherry	*Prunus*
Bird	*P. padus*
Fuji	*P. incisa*
Sour	*P. cerasus*
Yoshino	*P. yedoensis*
Cherry Laurel	*P. laurocerasus*
Cherry Pie	*Heliotropium arborescens*
Cherry Plum	*Prunus cerasifera*
Chervil	*Anthriscus cerefolium*
Bulbous	*Chaerophyllum bulbosum*
Salad	*Anthriscus cerefolium*
Wild	*Cryptoraenia*
Chestnut, Sweet or Spanish	*Castanea sativa*
Chestnut Vine	*Tetrastigma voinierianum*

Chick Pea	*Cicer arietinum*
Chickweed	*Stellaria*
Mouse-ear	*Cerastium*
Chicory	*Cichorium intybus*
Chile Pine	*Araucaria araucana*
Chilean Bellflower	*Lapageria rosea, Nolana*
Chilean Crocus	*Tecophilaea cyanocrocus*
Chilean Hazel	*Gevuina avellana*
Chilean Jasmine	*Mandevilla suaveolens*
China Aster	*Callistephus chinensis*
Chinaberry	*Melia azederach*
Chincherinchee	*Ornithogalum thyrsoides*
Chinese Cedar	*Cedrela sinensis*
Chinese Evergreen	*Aglaonema modesta*
Chinese Foxglove	*Rehmannia elata*
Chinese Gooseberry	*Actinidia chinensis*
Chinese-hat Plant	*Holmskioldia sanguinea*
Chinese Houses	*Collinsia heterophylla*
Chinese Jade	*Crassula arborescens*
Chinese Lantern	*Physalis alkekengi, Sandersonia aurantiaca*
Chives	*Allium schoenoprasum*
Chinese	*A. tuberosum*
Chocktaw root	*Apocynum cannabinum*
Chokeberry	*Aronia*
Black	*A. melanocarpa*
Red	*A. arbutifolia*
Cholla	*Opuntia*
Christmas Box	*Sarcococca*
Christmas Cheer	*Sedum* x *rubrotinctum*
Christmas Jewels	*Aechmea racineae*
Christmas Pride	*Ruellia macrantha*
Christmas Rose	*Helleborus niger*
Chufa	*Cyperus esculentus*
Cider Gum	*Eucalyptus gunnii*
Cigar Plant	*Cuphea ignea*
Cinderella Slippers	*Sinningia regina*
Cinquefoil	*Potentilla*
Citrange	*Citroncirus webberi*
Citron	*Citrus medica*
Citronella	*Collinsonia*
Clary	*Salvia sclarea*
Cleavers	*Galium aparine*
Cliff Brake, Green	*Pellaea viridis*
Purple	*P. atropurpurea*
Clock Vine	*Thunbergia grandiflora*
Clover	*Trifolium*

Alsike	*T. hybridum*
Hop	*T. agrarium*
White	*T. repens*
Cobnut	*Corylus avellana*
Cockscomb	*Celosia cristata*
Cockspur Thorn	*Crataegus crus-galli*
Coconut	*Cocos nucifera*
Double	*Lodoicea maldavica*
Coffee	*Coffea*
Wild	*Psychotria*
Cohosh	*Actaea*
Black	*Cimicifuga*
Blue	*Caulophyllum thalictroides*
Coltsfoot	*Tussilago*
Alpine	*Homogyne alpina*
Colombia Buttercup	*Oncidium cheirophorum*
Columbine	*Aquilegia*
Comfrey	*Symphytum*
Russian	*S. uplandicum*
Compass-plant	*Silphium laciniatum*
Cone Flower	*Rudbeckia*
Coral Berry	*Aechmea fulgens*
Coral Drops	*Bessera elegans*
Coral Plant	*Berberidopsis corallina, Russellia equisetiformis*
Coral Tree	*Erythrina crista-galli*
Coral Vine	*Antigonon leptopus*
Coriander	*Coriandrum sativum*
Corkscrew Rush	*Juncus effusus spiralis*
Cork Tree	*Phellodendron*
Corn	*Triticum*
(America)	*Zea mays*
Corn Cockle	*Agrostemma githago*
Corn Lily	*Ixia*
Corn Marigold	*Chrysanthemum segetum*
Corn Salad	*Valerianella locusta*
Cornel	*Cornus*
Cornelian Cherry	*Cornus mas*
Cornflower	*Centaurea cyanus*
Costmary	*Tanacetum balsamita*
Cotton	*Gossypium*
Levant	*G. herbaceum*
Tree	*G. arboreum*
Cottongrass	*Eriphorium*
Cotton Rose	*Hibiscus mutabilis*
Cottonwood	*Populus*
Cowbane	*Cicuta virosa*

221

Cow Herb	*Vaccaria pyramidata*
Cowberry	*Vaccinium vitis-idaea*
Cowpea	*Vigna sinensis*
Cowslip	*Primula veris*
Crab-apple	*Malus*
Siberian	*M. baccata*
Cranberry	*Oxycoccus*
American	*O. macrocarpus*
Small	*O. palustris*
Crane Flower	*Strelitzia reginae*
Cranesbill	*Geranium*
Bloody	*G. sanguineum*
Dusky	*G. phaeum*
Meadow	*G. pratense*
Wood	*G. sylvaticum*
Crape Myrtle	*Lagerstroemia indica*
Cream Cups	*Platystemon californicus*
Creeping Charlie	*Pilea nummularia*
Creeping Jenny	*Lysimachia nummularia*
Creeping Wintergreen	*Gaultheria procumbens*
Creeping Zinnia	*Sanvitalia procumbens*
Crocus, Autumn	*Colchicum autumnale*
Cross Vine	*Bignonia capreolata*
Cress, American	*Barbarea*
Bitter	*Cardamine*
Garden	*Lepidium sativum*
Indian	*Tropaeolum majus*
Stone	*Aethionema*
Winter	*Barbarea*
Crocus	*Crocus*
Saffron	*C. sativum*
Crosswort	*Crucciata laevipes*
Croton	*Codiaeum variegatum*
Crowberry	*Empetrum nigrum*
Crown Daisy	*Chrysanthemum coronarium*
Crown of Thorns	*Euphorbia milii*
Cruel Plant	*Araujia sericofera*
Cuckoo Flower	*Cardamine pratensis*
Cuckoo Pint	*Arum maculatum*
Cucumber	*Cucumis sativus*
Cucumber Tree	*Magnolia acuminata*
Cucuzzi	*Lagenaria siceraria*
Cudweed	*Gnaphalium*
Cup Flower	*Nierembergia*
Cupid's Dart	*Catananche caerulea*
Currant	*Ribes*
Black	*R. nigrum*

Buffalo	*R. odoratum*
Flowering	*R. sanguineum*
Golden	*R, aureum*
Mountain	*R. alpinum*
Red	*R. sylvestre*
Curry Plant	*Helichrysum italicum serotinum*
Custard Apple	*Annona cherimola, A. reticulata*
Cut-leaved Bramble	*Rubus laciniatus*
Cypress	*Cupressus*
Bald	*Taxodium distichum*
False	*Chamaecyparis*
Hinoki	*C. obtusa*
Italian	*Cupressus sempervirens*
Lawson	*Chamaecyparis lawsoniana*
Leyland	*Cupressocyparis leylandii*
Monterey	*Cupressus macrocarpa*
Nootka	*Chamaecyparis nootkatensis*
Pond	*Taxodium ascendens*
Sawara	*Chamaecyparis pisifera*
Cypress Spurge	*Euphorbia cyparissias*
Cypress Vine	*Ipomoea quamoclit*

D

Daffodil, Hoop-petticoat	*Narcissus bulbocodium*
Pheasant's Eye	*N. poeticus recurvus*
Wild	*N. pseudonarcissus*
Dahlia	*Dahlia*
Climbing	*Hidalgoa wercklei*
Daisy	*Bellis perennis*
Bush	*Olearia*
Michaelmas	*Aster*
Dame's Violet	*Hesperis matronalis*
Date Plum	*Diospyros lotus*
Dawn Redwood	*Metasequoia glyptostroboides*
Day Flower	*Commelina*
Day Lily	*Hemerocallis*
Dead Nettle	*Lamium*
Giant	*L. orvala*
Spotted	*L. maculatum*
Deodar	*Cedrus deodara*
Desert Privet	*Peperomia magnoliifolia*
Devil Flower	*Tacca chantrieri*
Devil's Backbone	*Kalanchoe daigremontiana*
Devil's Claw	*Physoplexis comosa*
Devil's Fig	*Argemone mexicana*
Devil's Ivy	*Epipremnum aureum*
Devil's Paintbrush	*Hieracium aurantiacum*
Devil's Tongue	*Amorphophallus rivieri*

Dill	*Anethum graveolens*
Dog's-tooth Violet	*Erythronium dens-canis*
Dogwood	*Cornus*
Common	*C. sanguinea*
Flowering	*C. florida*
Pacific	*C. nuttallii*
Douglas Fir	*Pseudotsuga menziesii*
Dove Tree	*Davidia involucrata*
Dragon Arum	*Dracunculus vulgaris*
Dragon Tree	*Dracaena draco*
Dropwort	*Filipendula vulgaris*
Drunkard's Dream	*Hatiora salicornioides*
Dumb Cane	*Dieffenbachia*
Durian	*Durio zibethinus*
Dusty Miller	*Artemisia stelleriana*
Dutchman's Breeches	*Dicentra cucullaris*
Dutchman's Pipe	*Aristolochia macrophylla*
Dwale	*Atropa bella-donna*
Dyer's Greenweed	*Genista tinctoria*

E

Earth Star	*Cryptanthus*
Earthnut	*Arachnis hypogaea*
Easter Lily	*Lilium longiflorum*
Eastern Red Cedar	*Juniperus virginiana*
Egg Plant	*Solanum melongena*
Eglantine	*Rosa eglanteria*
Egyptian Star Cluster	*Pentas lanceolata*
Elder	*Sambucus*
Common	*S. nigra*
Red-berried	*S. racemosa*
Elecampane	*Inula helenium*
Elephant's Ears	*Caladium*
Elephant's Foot	*Dioscorea elephantipes*
Elfin Herb	*Cuphea hyssopifolia*
Elk's Horns	*Rhombophyllum nelii*
Elm	*Ulmus*
Belgian	*U. belgic*
Camperdown	*U. camperdownii*
Cornish	*U. angustifolia cornubiensis*
Dutch	*U. hollandica*
English	*U. procera*
Exeter	*U. exoniensis*
Goodyer's	*U. angustifolia*
Jersey	*U. sarniensis*
Smooth	*U. carpinifolia*
Wheatley	*U. sarniensis*
Wych	*U. glabra*

Endive	*Cichorium endiva*
Evening Primrose	*Oenothera biennis*
Everlasting	*Helichrysum bracteatum*

F

Fair Maids of France	*Ranunculus aconitifolius flore pleno*
Fairy Bells	*Disporum*
Fairy Forget-me-not	*Eritrichium nanum*
Fairy Foxglove	*Erinus alpinus*
Fairy Moss	*Azolla*
False African Violet	*Streptocarpus saxorum*
False Aralia	*Dizygotheca elegantissima*
False Chamomile	*Boltonia*
False Goatsbeard	*Astilbe*
False Hellebore	*Veratrum*
False Spikenard	*Smilacina racemosa*
Fameflower	*Talinum*
Fanwort	*Cabomba*
Farkleberry	*Vaccinium arboreum*
Feather Grass	*Stipa pennata*
Felt Bush	*Kalanchoe beharensis*
Fennel	*Foeniculum vulgare*
Florence	*F. vulgare azoricum*
Giant	*Ferula communis*
Fenugreek	*Trigonella foenumgreacum*
Fern, Adder's tongue	*Ophioglossum*
Alpine Lady	*Athyrium distentifolium*
American Sword	*Polystichum munitum*
Arctic Bladder	*Cystopteris dickieana*
Asparagus	*Asparagus setosus*
Berry Bladder	*Cystopteris bulbifer*
Bird's Nest	*Asplenium nidus*
Bladder	*Cystopteris*
Boston	*Nephrolepis exaltata bostoniensis*
Brazil Tree	*Blechnum brasiliense*
Brittle Bladder	*Cystopteris fragilis*
Buckler	*Dryopteris*
Buckler, Broad	*D. dilatata*
Button	*Pellaea rotundifolia*
Christmas	*Polystichum acrostichoides*
Cinnamon	*Osmunda cinnamonea*
Crested Buckler	*Dryopteris cristata*
Crown	*Blechnum discolor*
Deer's Foot	*Davallia canariensis*
Elk's Horn	*Platycerium bifurcatum*
Erect Sword	*Nephrolepis cordifolia*
Filmy	*Hymenophyllum*
Floating	*Salvinia auriculata*

Giant Wood	*Dryopteris goldieana*
Golden Tree	*Dicksonia fibrosa*
Hairy Lip	*Cheilanthes lanosa*
Hammock	*Blechnum occidentale*
Hard	*B. spicant*
Hard Shield	*Polystichum aculeatum*
Hare's Foot	*Polypodium vulgare*
Hart's Tongue	*Phyllitis scolopendrium*
Hay-scented	*Dennstaedtia punctilobula*
Hay-scented Buckler	*Dryopteris aemula*
Hen and Chicken	*Asplenium bulbiferum*
Holly	*Polystichum acrostichoides*
Interrupted	*Osmunda claytoniana*
Japanese Painted	*Athyrium goeringianum pictum*
Lady	*A. filix-femina*
Lip	*Cheilanthes*
Maidenhair Fern	*Adiantum*
Male	*Dryopteris filix-mas*
Mountain Bladder	*Cystopteris montana*
Necklace	*Asplenium bulbiferum*
Oak	*Gymnocarpium dryopteris*
Ostrich-feather	*Matteuccia struthiopteris*
Palm Leaf	*Blechnum capense*
Parsley	*Cryptogramma crispa*
Rabbit's Foot	*Davallia fejeensis*
Rib	*Blechnum brasiliense*
Royal	*Osmunda regalis*
Rusty Back	*Ceterach officinarum*
Sensitive	*Onoclea sensibilis*
Soft Shield	*Polystichum setiferum*
Squirrel's Foot	*Davallia mariesii, D. trichomanoides*
Stag's Horn	*Platycerium bifurcatum*
Sword	*Nephrolepis*
Tree	*Cyathea*
Tunbridge Filmy Fern	*Hymenophyllum tunbrigense*
Walking	*Camptosorus rhizophyllus*
Wall	*Polypodium*
Wilson's Filmy	*Hymenophyllum wilsonii*
Woolly Rock	*Cheilanthes distans*
Woolly Tree	*Dicksonia antarctica*
Fiddler's Trumpets	*Sarracenia leucophylla*
Fig, Common	*Ficus carica*
Creeping	*F. pumila*
Mistletoe	*F. deltoidea*
Rusty	*F. rubiginosa*
Weeping	*F. benjamina*
Filbert	*Corylus maxima*

Finger Aralia	*Dizygotheca elegantissima*
Finger-nail Plant	*Neoregelia spectabilis*
Finocchio	*Foeniculum vulgare azoricum*
Fir	*Abies*
Alpine	*A. lasiocarpa*
Balsam	*A. balsamea*
Caucasian	*A. nordmanniana*
European Silver	*A. alba*
Flaky	*A. squamata*
Giant	*A. grandis*
Greek	*A. cephalonica*
Himalayan	*A. spectabilis*
Korean	*A. koreana*
Nikko	*A. homolepis*
Noble	*A. procera*
Pacific Silver	*A. amabilis*
Red	*A. magnifica*
Red Silver	*A. amabilis*
Santa Lucia	*A. bracteata*
White	*A. concolor*
Fire Bush	*Embothrium coccineum*
Fire-on-the-Mountain	*Euphorbia cyathophora*
Firecracker Flower	*Crossandra infundibuliformis, Dichelostemma ida-maia*
Firecracker Vine	*Manettia inflata*
Firethorn	*Pyracantha*
Firewheel Tree	*Stenocarpus sinuatus*
Five Fingers	*Syngonium auritum*
Flame Creeper	*Tropaeolum speciosum*
Flame Nettle	*Coleus blumei*
Flame of the Woods	*Ixora coccinea*
Flame Plant	*Anthurium scherzerianum*
Flame Violet	*Episcia cupreata*
Flaming Sword	*Vriesia splendens*
Flamingo Flower	*Anthurium scherzerianum*
Flax	*Linum usitatissimum*
Golden	*L. flavum*
Tree	*L. arboreum*
Yellow	*Reinwardtia indica*
Fleabane	*Erigeron*
Floss Flower	*Ageratum conyzoides*
Flower of an Hour	*Hibiscus trionum*
Flower of the Western Wind	*Zephyranthes candida*
Flowering Rush	*Butomus umbellatus*
Foam Flower	*Tiarella cordifolia*
Forest Lily	*Veltheimia viridiflora*
Forget-me-not	*Myosotis*

Fountain Grass	*Pennisetum setaceum*
Four o'clock	*Mirabilis jalapa*
Foxglove	*Digitalis purpurea*
Foxtail Grass	*Alopecurus pratensis*
Foxtail Lily	*Eremurus*
Frangipani	*Plumeria rubra*
Freckle Face	*Hypoestes phyllostachya*
Fremontia	*Fremontodendron*
French Bean	*Phaseolus vulgaris*
French Honeysuckle	*Hedysarum coronarium*
Friendship Plant	*Billbergia nutans, Pilea involucrata*
Fringe Tree	*Chionanthus virginicus*
Chinese	*C. retusus*
Frog's Bit	*Hydrocharis morsus-ranae*
Furze	*Ulex*

G

Gale	*Myrica gale*
Galingale	*Cyperus longus*
Garbanzo	*Cicer arietinum*
Gardener's Garters	*Phalaris arundinacea picta*
Garland Flower	*Daphne cneorum,*
	Hedychium coronarium
Garlic	*Allium sativum*
Crow	*A. vineale*
Gay Feather	*Liatris*
Gean	*Prunus avium*
German Ivy	*Senecio mikanioides*
Germander	*Teucrium*
Shrubby	*T. fruticans*
Wall	*T. chamaedrys*
Gherkin	*Cucumis sativus*
Ghost Plant	*Graptopetalum paraguayense*
Giant Caladium	*Alocasia cuprea*
Giant Elephant's Ear	*Alocasia*
Giant Hogweed	*Heracleum mantegazzianum*
Giant Reed	*Arundo donax*
Ginger	*Zingiber officinale*
Wild	*Asarum*
Ginger Lily	*Hedychium*
Scarlet	*H. coccineum*
Girasole	*Helianthus tuberosus*
Gladdon Iris	*Iris foetidissima*
Globe Amaranth	*Gomphrena globosa*
Globe Daisy	*Globularia*
Globe Thistle	*Echinops*
Globeflower	*Trollius*
Glory Bower	*Clerodendrum philippinum*

Glory Bush	*Tibouchina urvilleana*
Glory of Texas	*Thelocactus bicolor*
Glory Lily	*Gloriosa*
Glory Pea	*Clianthus puniceus*
Goat's Beard	*Aruncus dioicus*
Goat's Rue	*Galega officinalis*
Godetia	*Clarkia*
Gold Guinea	*Hibbertia scandens*
Golden-rayed Lily	*Lilium auratum*
Golden Alyssum	*Aurinia saxatilis*
Golden Bell	*Forsythia*
Golden Club	*Orontium*
Golden Column	*Trichocereus spachianus*
Golden Drop	*Onosma tauricum*
Golden Rod	*Solidago*
Golden Tom Thumb	*Parodia aureispina*
Golden Trumpet	*Allamanda cathartica*
Good King Henry	*Chenopodium bonushenricus*
Good Luck	*Cordyline terminalis*
Gooseberry	*Ribes uva-crispa*
Gorse	*Ulex*
Common	*U. europaeus*
Dwarf	*U. minor*
Spanish	*Genista hispanica*
Granadilla	*Passiflora edulis*
Giant	*P. quadrangularis*
Red	*P. coccinea*
Yellow	*P. laurifolia*
Grape, Fox	*Vitis labrusca*
Grape Hyacinth	*Muscari*
Oxford and Cambridge	*M. tubergenianum*
Grape Ivy	*Cissus rhombifolia*
Miniature	*C. striata*
Grape Vine, Common	*Vitis vinifera*
Grapefruit	*Citrus paradisi*
Grass of Parnassus	*Parnassia palustris*
Greater Spearwort	*Ranunculus lingua*
Grey Sage Brush	*Atriplex canescens*
Ground Ivy	*Glechoma hederacea*
Groundsel Tree	*Baccharis halimifolia*
Guava	*Psidium guajava*
Guelder Rose	*Viburnum opulus*
Guernsey Lily	*Nerine sarniensis*
Gumbo	*Abelmoschus esculentus*
Gumplant	*Grindelia*

H

Hackberry	*Celtis occidentalis*
Hair Grass	*Aira, Eleocharis acicularis*
Handkerchief Tree	*Davidia*
Harebell	*Campanula rotundifolia*
Hare's-tail Grass	*Lagurus ovatus*
Harlequin Flower	*Sparaxis tricolor*
Harry Lauder's Walking Stick	*Corylus avellana contorta*
Hawthorn	*Crataegus*
Common	*C. monogyna*
Midland	*C. laevigata*
Hazel	*Corylus avellana*
Turkish	*C. colurna*
Heart of Jesus	*Caladium*
Heartsease	*Viola tricolor*
Hearts on a String	*Ceropegia woodii*
Heath	*Erica*
Cornish	*E. vagans*
Cross-leaved	*E. tetralix*
Spanish	*E. australis*
Tree	*E. arborea*
Heather	*Calluna vulgaris*
Bell	*Erica cineria*
Hellebore	*Helleborus*
False	*Veratrum*
Green	*Helleborus viridis*
Stinking	*H. foetidus*
Hemlock	*Tsuga*
Eastern	*T. canadensis*
Mountain	*T. mertensiana*
Western	*T. heterophylla*
Hemp	*Cannabis sativa*
Hemp Agrimony	*Eupatorium cannabinum*
Henbane	*Hyoscyamus niger*
Henna	*Lawsonia inermis*
Herald's Trumpet	*Beaumontia grandiflora*
Herb Christopher	*Actaea spicata*
Hickory	*Carya*
Bitternut	*C. cordiformis*
Mockernut	*C. tomentosa*
Pignut	*C. glabra*
Shagbark	*C. ovata*
Holly	*Ilex*
Blue	*I. meserveae*
Common	*I. aquifolium*
Horned	*I. cornuta*
Hollyhock	*Alcea rosea*
Fig-leaved	*A. ficifolia*

Honeysuckle	*Lonicera*
Common	*L. periclymenum*
Perfoliate	*L. caprifolium*
Trumpet	*L. sempervirens*
Hop	*Humulus lupulus*
Hop Hornbeam	*Ostrya carpinifolia*
Hop Tree	*Ptelea trifoliata*
Hornbeam	*Carpinus*
American	*C. caroliniana*
Common	*C. betulus*
Hornwort	*Ceratophyllum*
Horse Briar	*Smilax rotundifolia*
Horse-chestnut	*Aesculus*
Common	*A. hippocastanum*
Indian	*A. indica*
Japanese	*A. turbinata*
Red	*A. carnea*
Horse-radish	*Armoracia rusticana*
Horseshoe Vetch	*Hippocrepis comosa*
Hottentot Fig	*Carpobrotus edulis*
Hot Water Plant	*Achimenes*
Hound's Tongue	*Cynoglossum officinale*
Houseleek	*Sempervivum*
Cobweb	*S. arachnoideum*
Common	*S. tectorum*
Humble Plant	*Mimosa pudica*
Huon Pine	*Dacrydium franklinii*
Hyacinth	*Hyacinthus orientalis*
Grape	*Muscari*

I

Ice Plant	*Hylotelephinum spectabile*
Immortelle	*Xeranthemum annuum*
Incense Cedar	*Calocedrus decurrens*
Inch Plant	*Callisia, Tradescantia albiflora*
Flowering	*Tradescantia blossfeldiana*
Striped	*Callisia elegans*
Indian Bean Tree	*Catalpa bignonioides*
Indian Currant	*Symphoricarpus rivularis*
Indian Hawthorn	*Rhaphiolepis indica*
Indian Physic	*Gillenia trifoliata*
Indian Shot	*Canna indica*
Indian Turnip	*Arisaema*
Indigo, False	*Amorpha fruticosa, Baptisia australis*
Wild	*Baptisia tinctoria*
Iris	*Iris*
English	*I. xiphoides*
Spanish	*I. xiphium*

Ironweed	*Vernonia*
Iron Wood	*Ostrya virginiana*
Ivy	*Hedera*
Common	*H. helix*
Irish	*H. hibernica*
Poet's	*H. helix poetica*
Ivy-leaved Toadflax	*Cymbalaria muralis*

J

Jack-in-the-Pulpit	*Arisaema triphyllum*
Jacob's Coat	*Acalypha wilkesiana*
Jacob's Ladder	*Pedilanthes tithymaloides smallii,*
	Polemonium caeruleum
Jade Plant	*Crassula argentea*
Japanese Foam Flower	*Tanakaea radicans*
Japan Pepper	*Zanthoxylum piperitum*
Jasmine	*Jasminum*
Jelly Beans	*Sedum pachyphyllum*
Jerusalem Cherry	*Solanum pseudocapsicum*
False	*S. capsicastrum*
Jerusalem Cross	*Lychnis chalcedonica*
Jerusalem Sage	*Phlomis fruticosa*
Jerusalem Thorn	*Parkinsonia aculeata*
Jessamine	*Jasminum*
Jewel Weed	*Impatiens capensis*
Job's Tears	*Coix lacryma-jobi*
Joe-pye Weed	*Eupatorium*
Joseph's Coat	*Amaranthus tricolor*
Joy Weed	*Alternanthera*
Judas Tree	*Cercis siliquastrum*
Juniper	*Juniper*
Jupiter's Beard	*Anthyllis barba-jovis*

K

Kaffir Lily	*Clivia miniata, Schizostylis coccinea*
Kahili ginger	*Hedychium gardnerianum*
Kaki	*Diospyros kaki*
Kale, Ornamental	*Brassica oleracea acephala*
Kamila Tree	*Mallotus*
Kangaroo Apple	*Solanum aviculare*
Kangaroo Vine	*Cissus antarctica*
Kangaroo's Paw	*Anigozanthus*
Katsura Tree	*Cercidiphyllum japonicum*
Kauri Pine	*Agathis australis*
Kentia	*Howea*
Kentucky Coffee Tree	*Gymnocladus dioica*
Kidney Bean	*Phasoleus vulgaris*
King Cup	*Caltha palustris*
King of the Alps	*Eritrichium nanum*

King Plant	*Anoectochilus regalis*
King William Pine	*Athrotaxis selaginoides*
Kingfisher Daisy	*Felicia bergeriana*
King's Spear	*Asphodeline lutea*
Kiwi Fruit	*Actinidia deliciosa*
Kohl Rabi	*Brassica oleracea gongylodes*
Kowhai	*Sophora tetraptera*
Kris Plant	*Alocasia lindeniana*
Kumquat	*Fortunella japonica*

L

Labrador Tea	*Ledum groenlandicum*
Lace Flower Vine	*Episcia dianthiflora*
Lace Trumpets	*Sarracenia leucophylla*
Ladies' Fingers	*Abelmoschus esculentus*
Lady of the Night	*Brassavola nodosa, Brunfelsia americana*
Lady's Mantle	*Alchemilla*
Alpine	*A. alpina*
Lady's Smock	*Cardamine pratensis*
Lamb's Lettuce	*Valerianella locusta*
Lamb's Tongue	*Stachys byzantina*
Larch	*Larix*
Common, European	*L. decidua*
Dunkeld	*L. eurolepsis*
Golden	*Pseudolarix amabilis*
Japanese	*Larix kaempferi*
Larkspur	*Consolida*
Laurustinus	*Viburnum tinus*
Lavender	*Lavandula*
Common	*L. angustifolia*
French	*L. stoechas*
Sea	*Limonium*
Lavender Cotton	*Santolina chamaecyparissus*
Leadwort	*Plumbago*
Leek	*Allium porrum*
Lemon	*Citrus limon*
Lemon Balm	*Melissa officinalis*
Lemon Mint	*Monarda citriodora*
Lemon-scented Gum	*Eucalyptus citriodora*
Lemon verbena	*Aloysia triphylla*
Lenten Rose	*Helleborus orientalis*
Lentil	*Lens culinaris*
Leopard Lily	*Belamcanda chinensis*
Leopard's Bane	*Arnica montana, Doronicum*
Lettuce	*Lactuca sativa*
Licorice	*Glycyrrhiza glabra*
Wild	*Abrus precatorius*

Lilac	*Syringa*
Common	*S. vulgaris*
Persian	*S. persica*
Rouen	*S. chinensis*
Lily of China	*Rohdea japonica*
Lily of the Palace	*Hippeastrum aulicum*
Lily of the Valley	*Convallaria majalis*
Lily of the Valley Tree	*Clethra arborea*
Lily Tree	*Magnolia denudata*
Lime	*Citrus aurantiifolia, Tilia*
Broad-leaved	*Tilia platyphyllos*
Common	*T. europaea*
European White	*T. tomentosa*
Red-twigged	*T. platyphyllos rubra*
Small-leaved	*T. cordata*
Ling	*Calluna vulgaris*
Lion's Ear	*Leonotis leonurus*
Lipstick Vine	*Aeschynanthus radicans*
Little Candles	*Mammillaria prolifera*
Living Stones	*Lithops*
Livingstone Daisy	*Dorotheanthus bellidiformis*
Locust	*Robinia pseudacacia*
Caspian	*R. caspica*
Honey	*Gleditsia triacanthos*
Loganberry	*Rubus loganobaccus*
Lollipop Plant	*Pachystachys lutea*
London Pride	*Saxifraga urbium*
Loofah Gourd	*Luffa aegyptiaca*
Loosestrife	*Lysimachia*
Purple	*Lythrum salicaria*
Yellow	*Lysimachia vulgaris*
Loquat	*Eriobotrya japonica*
Lords and Ladies	*Arum maculatum*
Lotus	*Nelumbo*
American	*N. lutea*
Blue Egyptian	*N. caerulea*
Indian or Sacred	*N. nucifera*
White Egyptian	*N. lotus*
Lovage	*Levisticum officinale*
Love Grass	*Eragrostis elegans*
Japanese	*E. amabilis*
Love-in-a-Mist	*Nigella damascena*
Love-in-a-Puff	*Cardiospermum halicacabum*
Love-lies-Bleeding	*Amaranthus caudatus*
Lungwort	*Pulmonaria*
Lyme Grass	*Elymus areanarius*

M Madagascar Jasmine — *Stephanotis floribunda*
Madagascar Lace Plant — *Aponogeton madagascariensis*
Madder — *Rubia tinctoria*
Madonna Lily — *Lilium candidum*
Madrona — *Arbutus menziesii*
Madwort — *Alyssum*
Mahogany Tree — *Swietenia mahogani*
Maidenhair Fern — *Adiantum*
   Australian — *A. formosum*
   Common — *A. capillus-veneris*
   Delta — *A. raddianum*
   Giant — *A. trapeziforme*
   Kashmir — *A. venustum*
   N. American — *A. pedatum*
   Rose-fronded — *A. pedatum japonicum*
   Rough — *A. hispidulum*
   Trailing — *A. caudatum*
Maidenhair Tree — *Ginkgo biloba*
Maize — *Zea mays*
Mallow — *Malva*
   Hairy — *Anisodontea scabrosa*
   Marsh — *Althaea officinalis*
   Musk — *Malva moschata*
   Poppy — *Callirhoe*
   Tree — *Lavatera arborea*
Maltese Cross — *Lychnis chalcedonica*
Mandarin Lime — *Citrus limonia*
Mandarin Orange — *C. reticulata*
Mandrake — *Mandragora*
   (U.S.A.) — *Podophyllum peltatum*
Mango — *Mangifera indica*
Mangosteen — *Garcinia mangostana*
Manuka — *Leptospermum scoparium*
Manzanita — *Arctostaphylos manzanita*
Maple — *Acer*
   Amur — *A. ginnala*
   Ash-leaved — *A. negundo*
   Field — *A. campestre*
   Hedge — *A. campestre*
   Hornbeam — *A. carpinifolium*
   Japanese — *A. palmatum*
   Montpellier — *A. monspessulanum*
   Nikko — *A. nikoense*
   Norway — *A. platanoides*
   Oregon — *A. macrophyllum*
   Red — *A. rubrum*
   Silver — *A. saccharinum*

Sugar	*A. saccharum*
Vine	*A. circinatum*
Marble Plant	*Neoregelia marmorata*
Mare's Tail	*Hippuris vulgaris*
Marguerite	*Chrysanthemum leucanthemum*
White	*C. frutescens*
Marigold	*Tagetes*
African	*T. erecta*
French	*T. patula*
Pot	*Calendula officinalis*
Signet	*Tagetes tenuifolia*
Mariposa Lily	*Calochortus*
Marjoram, Pot	*Origanum onites*
Sweet	*O. marjorana*
Wild	*O. vulgare*
Marmalade Bush	*Streptosolen jamesonii*
Marram Grass	*Ammophila arenaria*
Marrow	*Cucurbita pepo*
Marsh Marigold	*Caltha palustris*
Marvel of Peru	*Mirabilis jalapa*
Mask Flower	*Alonsoa*
Masterwort	*Astrantia*
May Apple	*Passiflora incarnata*
May Lily	*Maianthemum*
Maypop	*Passiflora incarnata*
Meadow Rue	*Thalictrum*
Meadowsweet	*Filipendula ulmaria*
Medlar	*Mespilus germanica*
Medusa's Head	*Euphorbia caput-medusae*
Melon	*Cucumis melo*
Bitter	*Momordica charantia*
Mercury, Dog's	*Mercuralis perennis*
Merrybells	*Uvularia*
Mescal Button	*Lophophora williamsii*
Mesquite	*Prosopis*
Mexican Orange Blossom	*Choisya ternata*
Mexican Sunflower	*Tithonia rotundifolia*
Mezereon	*Daphne mezereum*
Michaelmas Daisy	*Aster novi-belgii*
Mignonette	*Reseda odorata*
Milkweed	*Asclepias*
Millet	*Panicum miliaceum*
Mimosa	*Acacia dealbata*
Pink	*Albizia julibrissin*
Mind-your-own-Business	*Soleirolia soleirolii*
Mint	*Mentha*
Bowles'	*M. villosa alopecuroides*

Corsican	*M. requienii*
Curly	*M. piperita crispa*
Eau de Cologne	*M. piperita citrata*
Garden	*M. spicata*
Ginger	*M. gentilis*
Horse	*M. longifolia*
Round-leaved	*M. suaveolens*
Water	*M. aquatica*
Mint Bush	*Prostanthera*
Mistflower	*Eupatorium coelestinum*
Mistletoe	*Viscum album*
Mock Orange	*Philadelphus coronarius*
Mole Plant	*Euphorbia lathyris*
Monarch of the East	*Sauromatum venosum*
Monarch of the Veldt	*Venidium fastuosum*
Monkey Flower	*Mimulus*
Monkey Puzzle	*Araucaria araucana*
Monkshood	*Aconitum*
Montbretia	*Crocosmia*
Moonstones	*Pachyphytum oviferum*
Sticky	*P. glutinicaule*
Moosewood	*Acer pensylvanicum*
Mop-headed Acacia	*Robinia pseudacacia umbraculifera*
Morning Glory	*Ipomoea*
Mosaic Plant	*Fittonia verschaffeltii argyroneura*
Mossfern	*Selaginella pallescens*
Mother-in-law's Tongue	*Sansevieria trifasciata*
Mother of Pearl-Plant	*Graptophyllum paraguayense*
Mother of Thousands	*Kalanchoe daigremontiana, Saxifraga stolonifera*
Mount Atlas Daisy	*Anacyclus depressus*
Mount Wellington Peppermint	*Eucalyptus coccifera*
Mountain Ash	*Sorbus aucuparia*
Mountain Avens	*Dryas octopetala*
Mountain Laurel	*Kalmia latifolia*
Mountain Pepper	*Drimys lanceolata*
Mountain Tobacco	*Arnica montana*
Mourning Widow	*Geranium phaeum*
Mouse-tail Plant	*Arisarum proboscideum*
Moutan	*Paeonia suffruticosa*
Mrs Robb's Bonnet	*Euphorbia robbiae*
Mugwort	*Artemesia vulgaris*
Mulberry	*Morus*
Common	*M. nigra*
Paper	*Broussonetia papyrifera*
White	*Morus alba*
Mullein	*Verbascum*

Cretan	*V. creticum*
Dark	*V. nigrum*
Moth	*V. blattaria*
Nettle-leaved	*V. chaixii*
Purple	*V. phoeniceum*
Mung Bean	*Phaseolus aureus*
Myrtle	*Myrtus*
Common	*M. communis*
Tarentum	*M. communis tarentina*
Myrobalan	*Prunus cerasifera*

N

Nasturtium	*Tropaeolum majus*
Native's Comb	*Pachycereus pecten-aboriginum*
Navelwort	*Omphalodes*
Venus's	*O. linifolia*
Nettle Tree	*Celtis*
Never-never Plant	*Ctenanthe oppenheimiana*
New Zealand Burr	*Acaena*
New Zealand Daisy	*Celmisia*
New Zealand Flax	*Phormium tenax*
New Zealand Lilac	*Hebe hulkeana*
Nightshade, Deadly	*Atropa bella-donna*
Woody	*Solanum dulcamara*
Ninebark	*Physocarpus*
Nirre	*Nothofagus antarctica*
Norfolk Island Pine	*Araucaria heterophylla*

O

Oak	*Quercus*
Black	*Q. velutina*
Black Jack	*Q. marilandica*
Common	*Q. robur*
Cork	*Q. suber*
Daimio	*Q. dentata*
Durmast	*Q. petraea*
English	*Q robur*
Golden, of Cyprus	*Q. alnifolia*
Holm	*Q. ilex*
Hungarian	*Q. frainetto*
Lebanon	*Q. libani*
Lucombe	*Q. hispanica lucombeana*
Pedunculate	*Q. robur*
Pin	*Q. palustris*
Red	*Q. rubra*
Scarlet	*Q. coccinea*
Sessile	*Q. petraea*
Shingle	*Q. imbricaria*
Turkey	*Q. cerris*

Oat Grass	*Arrhenatherum elatius*
Oats	*Avena*
Obedient Plant	*Physostegia virginiana*
Ocean Spray	*Holodiscus discolor*
Oconee Bells	*Shortia galacifolia*
Okra	*Abelmoschus esculentus*
Old Maid	*Catharanthus roseus*
Old Man's Beard	*Clematis vitalba*
Old Woman	*Artemisia stelleriana*
Oleander	*Nerium oleander*
Olive	*Olea europaea*
Russian	*Elaeagnus*
Onion	*Allium cepa*
Welsh	*A. fistulosum*
Orach	*Atriplex hortensis*
Orange	*Citrus*
Bitter	*C. aurantium*
Seville	*C. aurantium*
Sweet	*C. sinensis*
Orange Root	*Hydrastis canadensis*
Orchid, Bee	*Ophrys apifera*
Black	*Coelogyne pandurata*
Butterfly	*Oncidium papilio*
Common Spotted	*Dactylorhiza fuchsii*
Cradle	*Anguloa*
Dancing Doll	*Oncidium flexuosum*
Early Purple	*Orchis mascula*
Fox-tail	*Aerides*
Lace	*Odontoglossum crispum*
Lady's Slipper	*Cypripedium*
Lily of the Valley	*Odontoglossum pulchellum*
Meadow	*Dactylorhiza incarnata*
Moth	*Phalaenopsis*
Pansy	*Miltonia*
Ram's Head Lady's Slipper	*Cypripedium arietinum*
Slipper	*Paphiopedilum*
Soldier	*Orchis militaris*
Star of Bethlehem	*Angraecum sesquipedale*
Tiger	*Odontoglossum grande*
Orchid Bush	*Bauhinia acuminata*
Orchid Tree, Purple	*B. variegata*
Oregon Grape	*Mahonia aquifolium*
Orris	*Iris germanica florentina*
Osage Orange	*Maclura pomifera*
Osier, Common	*Salix viminalis*
Purple	*S. purpurea*
Oso Berry	*Osmaronia cerasiformis*

Oswego Tea	*Monarda didyma*
Our Lady's Milk Thistle	*Silybum marianum*
Our Lord's Candle	*Yucca whipplei*
Oxlip	*Primula elatior*
Ox-tongue Lily	*Haemanthus coccineus*

P

Paeony	*Paeonia*
Pagoda Tree	*Plumeria rubra*
Paigle	*Primula veris*
Painted Daisy	*Chrysanthemum carinatum*
Painted Drop Tongue	*Aglaonema crispum*
Painted Feather	*Vriesia carinata*
Dwarf	*V. psittacina*
Painted Tongue	*Salpiglossis sinuata*
Painted Wood-lily	*Trillium undulatum*
Painter's Palette	*Anthurium andreanum*
Palm, Australian Fan	*Livistona australis*
Bamboo	*Chamaedorea erumpens*
Betel Nut	*Areca catechu*
Burmese Fishtail	*Caryota mitis*
Canary Island Date	*Phoenix canariensis*
Chinese Fan	*Livistona chinensis*
Chusan	*Trachycarpus fortunei*
Curly Sentry	*Howea belmoreana*
Date	*Phoenix dactylifera*
Desert Fan	*Washingtonia filifera*
Dwarf Fan	*Chamaerops humilis*
Fan	*Trachycarpus fortunei*
Fishtail	*Caryota*
Lady	*Rhapis*
Miniature Date	*Phoenix roebelinii*
Paradise	*Howea forsteriana*
Parlour	*Chamaedorea elegans*
Sago	*Cycas revoluta*
Sentry	*Howea*
Thread	*Washingtonia robusta*
Toddy	*Caryota urens*
Weddell	*Microcoelum weddellianum*
Wine	*Caryota urens*
Yatay	*Butia yatay*
Yellow	*Chrysalidocarpus lutescens*
Palmetto	*Sabal*
Pampas Grass	*Cortaderia*
Panamigo	*Pilea involucrata*
Panda Plant	*Kalanchoe tomentosa, Philodendron bipennifolium*
Pansy	*Viola*

Garden	*V. wittrockiana*
Panther Lily	*Lilium pardalinum*
Papyrus	*Cyperus papyrus*
Paris Daisy	*Chrysanthemum frutescens*
Parsley	*Petroselinum crispum*
Parsley Vine	*Vitis vinifera apiifolia*
Parsnip	*Pastinaca sativa*
Cow	*Heracleum Sphondilium*
Partridge Berry	*Mitchella repens*
Pasque Flower	*Pulsatilla*
Passion Flower	*Passiflora*
Blue	*P. caerulea*
Pawpaw	*Asimina triloba*
Pea, Bush	*Pultenaea*
Garden	*Pisum sativum*
Peanut	*Arachis hypogaea*
Pea Tree	*Caragana arborescens*
Peach	*Prunus persica*
Peacock Plant	*Calathea makoyana*
Peacock Tiger Flower	*Tigridia pavonia*
Pear	*Pyrus*
Common	*P. communis*
Willow-leaved	*P. salicifolia*
Pearl Fruit	*Margyricarpus pinnatus*
Pearl Grass	*Briza maxima*
Pearl Plant	*Haworthia margaritifera*
Pearlwort	*Sagina*
Pearly Everlasting	*Anaphalis*
Pebble plants	*Lithops*
Pecan	*Carya illinoensis*
Pelican Flower	*Aristolochia grandiflora*
Pennyroyal	*Mentha pulegium*
Pen Wiper	*Kalanchoe marmorata*
Pepper	*Piper*
Chilli	*Capsicum frutescens*
Christmas	*C. annuum*
Sweet	*C. annuum*
Pepper-and-salt	*Erigenia bulbosa*
Peppergrass	*Lepidium sativum*
Peppermint, Black	*Mentha piperita*
White	*M. piperita officinalis*
Periwinkle	*Vinca*
Greater	*V. major*
Lesser	*V. minor*
Madagascar	*Catharanthus roseus*
Persian Ironwood	*Parrotia persica*
Persian Shield	*Strobilanthes dyerianus*

Persian Violet	*Exacum affine*
Persimmon	*Diospyros virginiana*
Chinese	*D. kaki*
Peyote	*Lophophora williamsii*
Pheasant's Eye	*Adonis autumnalis*
Pickerel Weed	*Pontederia cordata*
Picotee	*Dianthus caryophyllus*
Piggy-back Plant	*Tolmiea menziesii*
Pimpernel, Bog	*Anagallis tenella*
Scarlet	*A. arvensis*
Pincushion Flower	*Hakea laurina*
Pine	*Pinus*
Aleppo	*P. halepensis*
Arolla	*P. cembra*
Austrian	*P. nigra*
Balkan	*P. peuce*
Beach	*P. contorta*
Bhutan	*P. wallichiana*
Big-cone	*P. coulteri*
Bishop	*P. muricata*
Bosnian	*P. heldreichii leucodermis*
Bristle-cone	*P. aristata*
Chinese	*P. tabuliformis*
Chinese White	*P. armandii*
Corsican	*P. nigra maritima*
Digger	*P. sabiniana*
Dwarf Mountain	*P. mugo*
Eastern White	*P. strobus*
Himalayan	*P. wallichiana*
Japanese Black	*P. thunbergii*
Japanese Red	*P. densiflora*
Japanese White	*P. parviflora*
Knobcone	*P. attenuata*
Lacebark	*P. bungeana*
Limber	*P. flexilis*
Lodgepole	*P. contorta latifolia*
Macedonian	*P. peuce*
Maritime	*P. pinaster*
Mexican White	*P. ayacahuite*
Monterey	*P. radiata*
Northern Pitch	*P. rigida*
Scots	*P. sylvestris*
Stone	*P. pinea*
Sugar	*P. lambertiana*
Umbrella	*Sciadopitys*
Western White	*Pinus monticola*
Western Yellow	*P. ponderosa*

Weymouth	*P. strobus*
Whitebark	*P. albicaulis*
Yunnan	*P. yunnanensis*
Pineapple	*Ananas comosus*
Wild	*A. bracteatus*
Pineapple Weed	*Matricaria matricaroides*
Pink	*Dianthus*
Cheddar	*D. gratianopolitanus*
Clove	*D. caryophyllus*
Fringed	*D. superbus*
Glacier	*D. glacialis*
Indian	*D. chinensis*
Maiden	*D. deltoides*
Pink Sand Verbena	*Abronia umbellata*
Pink Siris	*Albizia julibrissin*
Pitcher Plant	*Nepenthes, Sarracenia*
California	*Darlingtonia californica*
Northern	*Sarracenia purpurea*
Yellow	*S. flava*
Plane	*Platanus*
London	*P. acerifolia*
Oriental	*P. orientalis*
Plantain	*Plantago*
Plantain Lily	*Hosta*
Plover Eggs	*Adromischus cooperi*
Plum	*Prunus domestica*
Beach	*P. maritima*
Plum Yew	*Cephalotaxus*
Plume Bush	*Calomeria amaranthoides*
Poached Egg Flower	*Limnanthes douglasii*
Poinsettia	*Euphorbia pulcherrima*
Annual	*E. cyathophora*
Poke Weed	*Phytolacca americana*
Policeman's Helmet	*Impatiens glandulifera*
Polka-dot Plant	*Hypoestes phyllostachya*
Polyanthus	*Primula polyantha*
Polypody, Common	*Polypodium vulgare*
Limestone	*Gymnocarpium robertianum*
Pomegranate	*Punica granatum*
Pond Cypress	*Taxodium ascendens*
Pondweed	*Potamogeton*
Poor Man's Orchid	*Schizanthus*
Poplar	*Populus*
Balsam	*P. balsamifera*
Black	*P. nigra*
Grey	*P. canescens*
Lombardy	*P. nigra italica*

White	*P. alba*
Poppy	*Papaver*
Alpine	*P. alpinum*
Arctic	*P. nudicaulis*
Californian	*Eschscholzia californica*
Celandine	*Stylophorum diphyllum*
Crested	*Argemone platyceras*
Harebell	*Meconopsis quintuplinervis*
Himalayan Blue	*M. betonicifolia*
Iceland	*Papaver nudicaule*
Mexican Tulip	*Hunnemannia fumariifolia*
Opium	*Papaver somniferum*
Plume	*Macleaya cordata*
Prickly	*Argemone mexicana*
Red Horned	*Glaucium corniculatum*
Spanish	*Papaver rupifragum*
Tree	*Romneya*
Tulip	*Papaver glaucum*
Welsh	*Meconopsis cambrica*
Yellow Horned	*Glaucium flavum*
Portugal Laurel	*Prunus lusitanica*
Potato	*Solanum tuberosum*
Potato Bean	*Apios americana*
Potato Onion	*Allium cepa aggregatum*
Potato, Sweet	*Ipmoea batatas*
Powder Puff	*Calliandra*
Prayer Plant	*Maranta leuconeura*
Prickly Moses	*Acacia verticillata*
Prickly Pear	*Opuntia*
Prickly Thrift	*Acantholimon*
Pride of Texas	*Phlox drummondii*
Primrose	*Primula vulgaris*
Fairy	*P. malacoides*
Prince Albert's Yew	*Saxegothaea conspicua*
Prince's Feather	*Amaranthus hybridus*
Princess of the Night	*Selenicereus pteranthus*
Princess Vine	*Cissus sicyoides*
Privet	*Ligustrum*
Common	*L. vulgare*
Propeller Plant	*Crassula falcata*
Prophet Flower	*Arnebia pulchra*
Pummelo	*Citrus maxima*
Pumpkin	*Cucurbita maxima*
Purple Bell Vine	*Rhodochiton volubile*
Purple Heart	*Setcreasia pallida*
Purple Moor Grass	*Molinia caerulea*
Purple Passion Vine	*Gynura aurantiaca*

Purple Top	*Verbena bonariensis*
Purple Wreath	*Petrea volubilis*
Purslane	*Portulaca oleracea*
Rock	*Calandrinia*
Tree	*Atriplex halimus*
Pussy Ears	*Cyanotis somaliensis, Kalanchoe tomentosa*
Pyrethrum, Common	*Chrysanthemum coccineum*
Dalmatian	*C. cinerariifolium*

Q

Quaking Grass	*Briza media*
Queen of the Night	*Selenicereus grandiflorus*
Queen of the Prairie	*Filipendula rubra*
Queen's Wreath	*Petrea volubilis*
Queensland Umbrella Tree	*Schefflera actinophylla*
Quillwort	*Isoetes*
Quince	*Cydonia oblonga*
Japanese	*Chaenomeles*

R

Rabbit's Foot	*Maranta leuconeura kerchoviana*
Radish	*Raphanus sativus*
Raffia	*Raphia*
Rainbow Star	*Cryptanthus bromelioides*
Rainbow Vine	*Pellionia pulchra*
Rampion	*Campanula rapunculus*
Spiked	*Phyteuma spicatum*
Ramsons	*Allium ursinum*
Rape	*Brassica napus*
Raspberry	*Rubus idaeus*
Rattan	*Calamus*
Rattlesnake Plant	*Calathea lancifolia*
Rauli	*Nothofagus procera*
Redbud	*Cercis canadensis*
Red Cape Tulip	*Haemanthus*
Red Ivy	*Hemigraphis alternata*
Red-hot Cat's Tail	*Acalypha hispida*
Red Hot Poker	*Kniphofia*
Red Nodding Bells	*Streptocarpus dunnii*
Red Ribbons	*Clarkia concinna*
Redwood, Coast	*Sequoia sempervirens*
Giant	*Sequoiadendron giganteum*
Reed Canary Grass	*Phalaris arundinacea*
Reed Grass	*Glyceria maxima*
Reed-mace	*Typha*
Resurrection Lily	*Lycoris squamigera*
Resurrection Plant	*Selaginella lepidophylla*
Rex-begonia Vine	*Cissus discolor*

Rhubarb	*Rheum rhabarbarum*
Ribbon Gum	*Eucalyptus viminalis*
Rice	*Oryza*
Roblé	*Nothofagus obliqua*
Rock Jasmine	*Androsace*
Rock Lily	*Arthropodium*
Rock Rose	*Cistus*
Rosary Vine	*Ceropegia woodii*
Rose	*Rosa*
Banksian	*R. banksiae*
Burnet	*R. pimpinellifolia*
Damask	*R. damascena*
Dog	*R. canina*
Himalayan Musk	*R. brunonii*
Holland	*R. centifolia*
Provence	*R. centifolia*
Threepenny-bit	*R. elegantula persetosa*
White	*R. x alba*
York and Lancaster	*R. damascena versicolor*
Rose Acacia	*Robinia hispida*
Rose of China	*Hibiscus rosa-sinensis*
Rose of Heaven	*Silene coeli-rosea*
Rose Pincushion	*Mammillaria zeilmanniana*
Rose of Sharon	*Hypericum calyciman*
Roseroot	*Rhodiola rosea*
Rough Bindweed	*Smilax aspera*
Rowan	*Sorbus aucuparia*
Royal Red Bugler	*Aeschynanthus pulcher*
Royal Paint Brush	*Haemanthus magnificus*
Royal Nodding Bells	*Streptocarpus wendlandii*
Rubber Plant	*Ficus elastica*
Rue	*Ruta graveolens*
Runner Bean	*Phaseolus vulgaris*
Scarlet	*P. coccineus*
Rupturewort	*Herniaria glabra*
Rutabaga	*Brassica napus napobrassica*

S

Safflower	*Carthamus tinctorius*
Saffron	*Crocus sativus*
False	*Carthamus tinctorius*
Saffron Spike	*Aphelandra squarrosa*
Sage	*Salvia*
Cardinal	*S. fulgens*
Common	*S. officinalis*
Gentian	*S. patens*
Mealy-cup	*S. farinacea*
Pineapple-scented	*S. rutilans*

Sage Brush	*Artemisia tridentata*
Saguaro	*Carnegiea gigantea*
Sainfoin	*Onobrychis viciifolia*
Salal	*Gaultheria shallon*
Sallow. Common	*Salix cinerea*
Great	*S. caprea*
Salmon Blood Lily	*Haemanthus multiflorus*
Salmonberry	*Rubus spectabilis*
Salsify	*Tragopogon porrifolius*
Black	*Scorzonera hispanica*
Salt Tree	*Halimodendron halodendron*
Saltwort	*Salsola*
Samphire, Marsh	*Salicornia*
Rock	*Crithmum maritimum*
Sandalwood	*Santalum*
Red	*Adenanthera pavonina*
Sand Myrtle	*Leiophyllum buxifolium*
Sandwort	*Arenaria*
Sapphire Flower	*Browallia speciosa*
Sassafras	*Sassafras albidum*
Satin Flower	*Clarkia amoena*
Satinwood Tree	*Murraya*
Satsuma	*Citrus reticulata*
Saucer Plant	*Aeonium undulatum*
Savin	*Juniperus sabina*
Savory, Summer	*Satureja hortensis*
Winter	*S. montana*
Scallion	*Allium cepa*
Scarborough Lily	*Vallota speciosa*
Scarlet Leadwort	*Plumbago indica*
Scarlet Trompetilla	*Bouvardia longiflora*
Scorpion senna	*Coronilla emerus*
Scotch Thistle	*Onopordum acanthium*
Screw Pine	*Pandanus*
Sea Buckthorn	*Hippophae rhamnoides*
Sea Campion	*Silene vulgaris maritima*
Sea Daffodil	*Pancratium maritimum*
Sea Holly	*Eryngium maritimum*
Sea Kale	*Crambe maritima*
Sea Lily	*Pancratium maritimum*
Sea Squill	*Urginea maritima*
Sea Urchin	*Hakea laurina*
Sedge	*Carex*
Great Pond	*C. riparia*
Umbrella	*Cyperus alternifolius*
Seersucker Plant	*Geogenanthus undatus*
Self Heal	*Prunella*

Senna	*Cassia*
American	*C. marylandica*
Sensitive Plant	*Mimosa pudica*
Service Tree	*Sorbus domestica*
Wild	*S. torminalis*
Serviceberry	*Amelanchier*
Seven Fingers	*Schefflera digitata*
Shaddock	*Citrus grandis*
Shallon	*Gaultheria shallon*
Shallot	*Allium cepa aggregatum*
Shamrock	*Oxalis acetosella*
(Ireland)	*Trifolium dubium*
(Ireland, U.S.A.)	*T. repens*
Shamrock Pea	*Parocheios communis*
Shasta Daisy	*Chrysanthemum* x *superbum*
Sheep Laurel	*Kalmia angustifolia*
Sheep's Bit	*Jasione*
Shell Flower	*Moluccella laevis*
Shepherd's Purse	*Capsella bursa-pastoris*
Shooting Star	*Dodecatheon*
Shrimp Plant	*Justicea brandegeana*
Siberian Squill	*Scilla sibirica*
Silk Tree	*Albizia julibrissin*
Silky Oak	*Grevillea robusta*
Silver Bell	*Halesia*
Silver Berry	*Elaeagnus commutata*
Silver Crown	*Cotyledon undulata*
Silver Gum	*Eucalyptus cordata*
Silver Squill	*Ledebouria socialis*
Silver Torch	*Cleistocactus straussii*
Silver Tree	*Leucadendron argenteum*
Silver Vine	*Actinidia polygama, Scindapsus pictus*
Skull-cap	*Scutellaria*
Skunk Cabbage	*Lysichiton americanum, Symplocarpus foetidus*
Slipper Flower	*Pedilanthes tithymaloides*
Slipperwort	*Calceolaria*
Sloe	*Prunus spinosa*
Small-leaved Gum	*Eucalyptus parvifolia*
Smilax (florists's)	*Asparagus asparagoides*
Smoke Tree	*Cotinus coggygria*
Snake Vine	*Hibbertia scandens*
Snake's-head Iris	*Hermodactylus tuberosus*
Snapdragon	*Antirrhinum majus*
Sneezeweed	*Helenium*
Sneezewort	*Achillea ptarmica*
Snow Gum	*Eucalyptus niphophila*

Tasmanian	*E. coccifera*
Snow-in-Summer	*Cerastium tomentosum*
Snow on the Mountain	*Euphorbia marginata*
Snowberry	*Symphoricarpos albus*
Snowdrop	*Galanthus*
Common	*G. nivalis*
Snowdrop Tree	*Halesia*
Mountain	*H. monticola*
Snowflake	*Leucojum*
Spring	*L. vernum*
Summer	*L. aestivum*
Snowy Mespilus	*Amelanchier ovalis*
Soapwort	*Saponaria officinalis*
Rock	*S. ocymoides*
Solomon's Seal	*Polygonatum*
Sorrel, Common	*Rumex acetosa*
French	*R. scutatus*
Sorrel Tree	*Oxydendrum arboreum*
Southern Beech	*Nothofagus*
Southernwood	*Artemisia abrotanum*
Sowbread	*Cyclamen*
Spanish Bayonet	*Yucca aloifolia*
Spanish Shawl	*Heterocentron elegans*
Spanish Moss	*Tillandsia usneoides*
Spatterdock	*Nuphar advena*
Spearmint	*Mentha spicata*
Speedwell	*Veronica*
Speedy Jenny	*Tradescantia fluminensis*
Spice Bush	*Lindera benzoin*
Spider Flower	*Cleome hassleriana*
Spider Lily	*Hymenocallis*
Golden	*Lycoris africana*
Red	*L. radiata*
Spider Plant	*Chlorophytum comosum*
Spike Heath	*Bruckenthalia spiculifolia*
Spikenard, American	*Aralia racemosa*
False	*Smilacina*
Spinach	*Spinacia oleracea*
Spinach Beet	*Beta vulgaris*
Spindle Tree	*Euonymus europaeus*
Spinning Gum	*Eucalyptus perriniana*
Spleenwort	*Asplenium*
Black	*A. adiantum-nigrum*
Ebony	*A. platyneuron*
Green	*A. viride*
Hanging	*A. flaccidum*
Maidenhair	*A. trichomanes*

Mother	*A. bulbiferum*
Sea	*A. marinum*
Spreading Clubmoss	*Selaginella kraussiana*
Spring Meadow Saffron	*Bulbocodium vernum*
Spruce	*Picea*
Black	*P. mariana*
Brewer	*P. breweriana*
Colorado	*P. pungens*
Dragon	*P. asperata*
Hondo	*P. jezoensis hondoensis*
Norway	*P. abies*
Serbian	*P. omorika*
Sitka	*P. sithensis*
White	*P. glauca*
Yezo	*P. yezoensis*
Spurge	*Euphorbia*
Spurge Laurel	*Daphne laureola*
Squirrel Tail Grass	*Hordeum jubatum*
Standing Cypress	*Ipomopsis rubra*
Star-flowered Lily of the Valley	*Smilacina stellata*
Star of Bethlehem	*Ornithogalum umbellatum*
Star of the Veldt	*Dimorphotheca sinuata*
Starfish Plant	*Cryptanthus acaulis, Stapelia variegata*
Stars of Persia	*Allium christophii*
Stinking Benjamin	*Trillium erectum*
Stock	*Matthiola*
Brompton	*M. incana*
Night-scented	*M. longipetala bicornis*
Ten Weeks	*M. incana Annua*
Virginia	*Malcolmia maritima*
Stone Cress	*Aethionema*
Stonecrop	*Sedum*
Biting	*S. acre*
Golden	*S. adolphi*
Strap Flower	*Anthurium crystallinum*
Strawberry	*Fragaria*
Alpine	*F. vesca*
Garden	*F. ananassa*
Hautbois	*F. moschata*
Mock	*Duchesnea indica*
Strawberry Tree	*Arbutus unedo*
String of Beads	*Senecio rowleyanus*
Sturt's Desert Pea	*Clianthus formosus*
Sugar Beet	*Beta vulgaris*
Sugarberry	*Celtis laevigata*
Sumach	*Rhus*

Smooth	*R. glabra*
Stag's Horn	*R. typhina*
Summer Cypress	*Kochia scoparia*
Summer Hyacinth	*Galtonia candicans*
Summer Torch	*Billbergia pyramidalis*
Sun Plant	*Portulaca grandiflora*
Sun Rose	*Helianthemum*
Sundew	*Drosera*
Sundrops	*Oenothera fruticosa*
Sunflower	*Helianthus annuus*
Swamp Cypress	*Taxodium distichum*
Swan River Daisy	*Brachycome iberidifolia*
Swede	*Brassica napus napobrassica*
Swedish Ivy	*Plectranthus australis*
Sweet Alyssum	*Lobularia maritima*
Sweet Bay	*Laurus nobilis*
Sweet Bergamot	*Monarda didyma*
Sweet Box	*Sarcococca*
Sweet Briar	*Rosa eglanteria*
Sweet Cicely	*Myrrhis odorata*
Sweet Fern	*Comptonia peregrina*
Sweet Flag	*Acorus calamus*
Sweet Four o'Clock Plant	*Mirabilis longiflora*
Sweet Gale	*Myrica gale*
Sweet Gum	*Liquidambar styraciflua*
Sweet Pea	*Lathyrus odoratus*
Sweet Pepper Bush	*Clethra alnifolia*
Sweet Potato Vine	*Ipomoea batatas*
Sweet Rocket	*Hesperis matronalis*
Sweet Sop	*Annona squamosa*
Sweet Sultan	*Centaurea moschata*
Sweet William	*Dianthus barbatus*
Sweet Woodruff	*Galium odoratum*
Swiss Chard	*Beta vulgaris*
Swiss Cheese Plant	*Monstera deliciosa*
Sycamore	*Acer pseudoplatanus*

T

Tail Flower	*Anthurium crystallinum*
Tailor's Patch	*Crassula lactea*
Tamarind	*Tamarindus indica*
Tamarisk	*Tamarix*
Tangerine	*Citrus reticulata*
Tansy	*Tanacetum vulgare*
Tarragon, French	*Artemisia dracunculus*
Russian	*A. dracunculus inodora*
Tasmanian Blue Gum	*Eucalyptus globulus*
Tassel Flower	*Emilia javanica*

Tassel Hyacinth	*Muscari comosum*
Tea Plant	*Camellia sinensis*
Tea Tree	*Leptospermum scoparium*
Teak Tree	*Tectona grandis*
Teasel	*Dipsacus fullonum*
Teddy Bear Plant	*Cyanotis kewensis*
Telegraph Plant	*Desmodium gyrans*
Temple Bells	*Smithiantha*
Three Birds Flying	*Linaria triornithophora*
Thrift	*Armeria*
Common	*A. maritima*
Jersey	*A. alliacea*
Throatwort	*Campanula trachelium, Trachelium caeruleum*
Tickseed	*Coreopsis*
Tiger Lily	*Lilium lancifolium*
Tiger's Jaws	*Faucaria tigrina*
Tingiringi Gum	*Eucalyptus glaucescens*
Toad Lily	*Tricyrtis hirta*
Toad Plant	*Stapelia variegata*
Toadflax	*Linaria*
Alpine	*L. alpina*
Common	*L. vulgaris*
Toadshade	*Trillium sessile*
Tobacco Plant	*Nicotiana tabacum*
Tomato	*Lycopersicum esculentum*
Cherry	*L. esculentum ceraisiforme*
Toothache Tree	*Zanthoxylum americanum*
Torch Lily	*Kniphofia*
Tortoise Plant	*Dioscorea elephantipes*
Touch-me-not	*Impatiens noli-tangere*
Trailing Arbutus	*Epigaea repens*
Trailing Watermelon Begonia	*Pellionia daveauana*
Transvaal Daisy	*Gerbera jamesonii*
Traveller's Joy	*Clematis vitalba*
Treasure Flower	*Gazania rigens*
Tree of Heaven	*Ailanthus altissima*
Triplet Lily	*Brodiaea coronaria*
Trout Lily	*Erythronium revolutum*
Trumpet Vine	*Campsis radicans*
Tuberose	*Polianthes tuberosa*
Tulip Tree	*Liriodendron tulipifera*
Chinese	*L. chinense*
Tupelo	*Nyssa sylvatica*
Turnip	*Brassica rapa*
Indian	*Arisaema*
Swede	*Brassica napo-brassica*

Turk's Cap Lily	*Lilium martagon*
Turtle Head	*Chelone*
Tutsan	*Hypericum androsaemum*
Twin Flower	*Linnaea borealis*

U

Umbrella Pine	*Sciadopitys verticillata*
Umbrella Plant	*Cyperus involucratus, Peltiphyllum peltatum*
Umbrella Tree	*Magnolia tripetala*
Ear-leaved	*M. fraseri*
Unicorn Plant	*Proboscidea louisianica*
Unicorn Root	*Aletris farinosa*
Upas Tree	*Antiaris toxicaria*
Urn Gum	*Eucalyptus urnigera*
Urn Plant	*Aechmea fasciata*

V

Valerian	*Valeriana*
Valerian, Red	*Centranthus ruber*
Vanilla	*Vanilla*
Velvet Leaf	*Kalanchoe beharensis*
Velvet Plant	*Gynura aurantiaca*
Training	*Ruellia makoyana*
Venus's Fly Trap	*Dionaea muscipula*
Venus's Looking Glass	*Legousia speculum-veneris*
Vervain	*Verbena*
Common	*V. officinalis*
Lilac	*V. rigida*
Rose	*V. x hybrida*
Violet	*Viola*
Australian	*V. hederacea*
Bird's-foot	*V. pedata*
Horned	*V. cornuta*
Marsh	*V. cucullata*
Olympian	*V. gracilis*
Sweet	*V. odorata*
Violet Cress	*Ionopsidium acaule*
Viper's Bugloss	*Echium vulgare*
Virginia Creeper	*Parthenocissus quinquefolia*
Virginia Cowslip	*Mertensia maritima*

W

Wake Robin	*Trillium grandiflorum*
Wall Rue	*Asplenium ruta-muraria*
Wallflower	*Cheiranthus cheiri*
Siberian	*C. allionii*
Walnut	*Juglans*
Black	*J. nigra*
Common	*J. regia*

English	*J. regia*
Japanese	*J. ailantifolia*
Texan	*J. microcarpa*
Wandflower	*Dierama pulcherrimum, Sparaxis*
Wandering Jew	*Tradescantia albiflora,*
	Zebrina pendula
Waratah	*Telopea speciosissima*
Tasmanian	*T. truncata*
Wart Plant	*Haworthia tesselata*
Water Avens	*Geum rivale*
Water Caltrop	*Trapa natans*
Water Chestnut	*Trapa natans*
Chinese	*Eleocharis dulcis*
Watercress	*Nasturtium*
Water Crowfoot	*Ranunculus aquatilis*
Water Figwort	*Scrophularia auriculata*
Water Hyacinth	*Eichhornia crassipes*
Water Lettuce	*Pistia stratiotes*
Water Lily, Cape Blue	*Nymphaea capensis*
Royal	*Victoria amazonica*
Santa Cruz	*V. cruziana*
White	*Nymphaea alba*
Yellow	*Nuphar lutea*
Water Melon	*Citrullus lanatus*
Water Milfoil	*Myriophyllum*
Water Plantain	*Alisma plantago-aquatica*
Water Poppy	*Hydrocleys nymphoides*
Water Soldier	*Stratiotes aloides*
Water Sprite	*Ceratopteris thalictroides*
Water Starwort	*Callitriche*
Water Trumpet	*Cryptocoryne*
Water Violet	*Hottonia palustris*
Watercress	*Nasturtium officinale*
Waterwillow	*Justicia americana*
Wattle	*Acacia*
Cootamundra	*A. baileyana*
Ovens	*A. pravissima*
Queensland	*A. podalyriifolia*
Rice's	*A. riceana*
Silver	*A. dealbata*
Sydney Golden	*A. longifolia*
Wax Flower	*Stephanotis floribunda*
Wax Myrtle	*Myrica cerifera*
Wax Plant	*Hoya carnosa*
Miniature	*H. bella*
Wax Privet	*Peperomia glabella*
Waxweed	*Cuphea*

Waxwork	*Celastrus scandens*
Wayfaring Tree	*Viburnum lantana*
Weaver's Broom	*Sparticum junceum*
Wedding Bush	*Ricinocarpos*
Wellingtonia	*Sequoiadendron giganteum*
Western Red Cedar	*Thuja plicata*
Western White Cedar	*T. occidentalis*
Wheat	*Triticum*
Whin	*Ulex*
White Paint-brush	*Haemanthus albiflos*
White Raintree	*Brunfelsia undulata*
White Snake-root	*Eupatorium rugosum*
White Velvet	*Tradescantia sillamontana*
Whitebeam	*Sorbus aria*
Swedish	*S. intermedia*
Whorl Flower	*Morina longifolia*
Whortleberry	*Vaccinium myrtillus*
Caucasian	*V. arctostaphylos*
Willow	*Salix*
Bay	*S. pentandra*
Crack	*S. fragilis*
Creeping	*S. repens*
Cricket Bat	*S. alba caerulea*
Dragon's Claw	*S. matsudana tortuosa*
Goat	*S. caprea*
Golden	*S. alba vitellina*
Musk	*S. aegyptiaca*
Peking	*S. matsudana*
Silver	*S. alba argentea*
Violet	*S. daphnoides*
Weeping	*S. babylonica*
White	*S. alba*
Woolly	*S. lanata*
Willow Herb	*Epilobium*
Winterberry	*Ilex verticillata*
Wintergreen	*Gaultheria procumbens*
Wineberry	*Rubus phoenicolasius*
Wingnut	*Pterocarya*
Caucasian	*P. fraxinifolia*
Winter Aconite	*Eranthis hyemalis*
Winter Cherry	*Solanum capsicastrum*
Winter Heliotrope	*Petasites fragrans*
Winter Sweet	*Chimonanthus praecox*
Wintergreen	*Pyrola*
Winter's Bark	*Drimys winteri*
Wishbone Flower	*Tovenia fournieri*
Witch Grass	*Panicum capillare*

Witch Hazel	*Hamamelis*
Chinese	*H. mollis*
Japanese	*H. japonica*
Withe-rod	*Viburnum cassinoides*
Woad	*Isatis tinctoria*
Wolf's Bane	*Aconitum vulparia*
Wonga-wonga Vine	*Pandorea jasminoide*
Wood Sorrel	*Oxalis acetosella*
Woodbine	*Lonicera periclymenum*
Wormwood	*Artemisia*
Common	*A. absinthium*
Sweet	*A. annua*

Y

Yam	*Dioscorea*
Ornamental	*D. discolor*
Yarrow	*Achillea millefolium*
Yellow Adder's Tongue	*Erythronium americanum*
Yellow Archangel	*Galeobdolon luteum*
Yellow Elder	*Tecoma stans*
Yellow Flag	*Iris pseudacorus*
Yellow Jessamine	*Gelsemium sempervirens*
Yellow Sage	*Lantana camara*
Yellow Wood	*Cladrastis sinensis*
Yellowroot	*Xanthorhiza simplicissima*
Yesterday Today & Tomorrow	*Brunfelsia calycina*
Yew	*Taxus*
Common	*T. baccata*
Irish	*T. baccata hibernica*
West Felton	*T. baccata dovastoniana*
Yorkshire Fog	*Holcus lanatus*
Youth & Old-age	*Aichryson domesticum*
Yulan	*Magnolia denudata*

Z

Zebra Basket Vine	*Aeschynanthus marmoratus*
Zebra Plant	*Aphelandra squarrosa,*
	Calathea zebrina, Cryptanthus zonalis